WEST CORK RAILWAYS

CHRIS LARKIN

MERCIER PRESS

Clontarf Bridge
An Indian ink drawing by Chris Larkin.

CONTENTS

Introduction	8
BIRTH	10
The Birth of the West Cork Railway	12
'Night Owls' Anti-Social Hours	38
Farming and the Railway	40
Animals	43
Tourism	47
Railway Bars	53
Railway Apparel	53
Fury on the Rails	54
Incidents	57
Travelogue	58
Drimoleague to Baltimore	76
BEAUTY	78
BETRAYAL	164
The Destruction of the Railway	167
A Green Future	175
Appendix 1 – Poems and Art	178
Appendix 2 – Railway Staff	186
Appendix 3 – Signalling, Carriages	192
Appendix 4 – Locomotives	198
Appendix 5 – Level crossings	210
Appendix 6 – Stations	214
Acknowledgments	223

This book is dedicated to Sheila, my mother and Patrick, my father, Patrica and Catherine, my sisters, Jerry, my brother, and all my nieces and nephews.

MAP OF WEST CORK RAILWAY SYSTEM

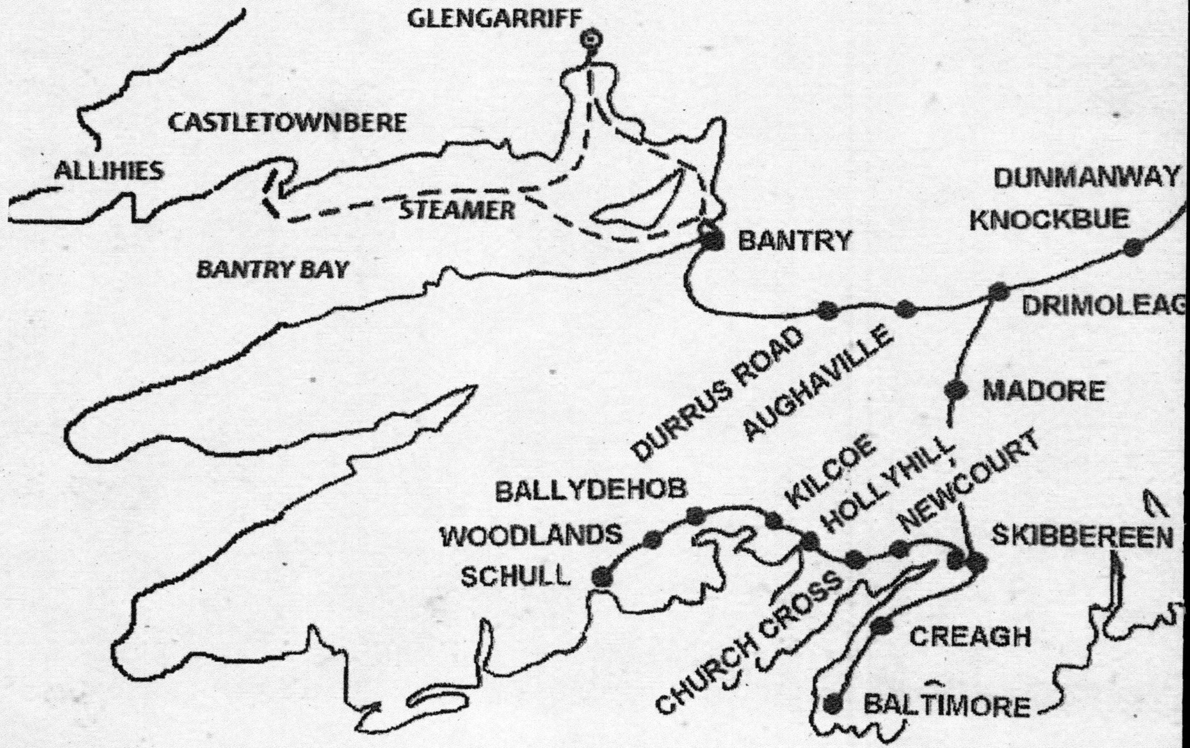

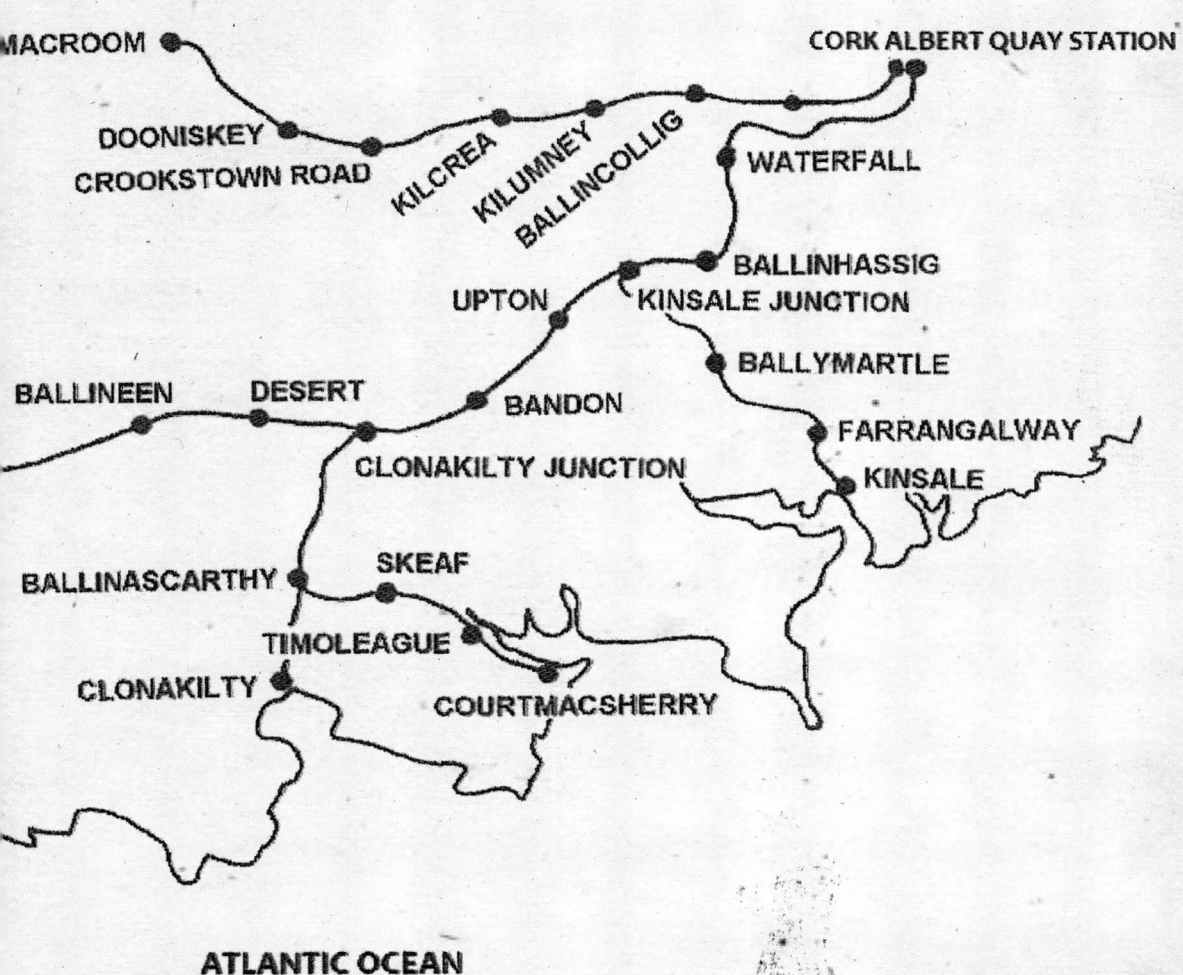

INTRODUCTION

'Die Eindrücke im elternhaus sind bestimmend fur das leiben eines menshen.'
A man's life is shaped by the first impressions received in the home of his parents.

Old German Proverb

Many years ago I became attracted to and interested in art, music, books, history, transport, politics, collie dogs, country life, sport and design. All of the above have been more than well served by very fine books, lectures, TV, films, etc. – with the notable exception of Irish railways.

My love of transport began at a young age. I grew up near that great Cork landmark, the Chetwynd Viaduct, just outside Bishopstown. My brother, Jerry and I used to watch the goods and passenger trains passing Milepost 4 (Milestones were used to pinpoint the exact location of a train at any given time and are still used today). We had a bird's eye view of the viaduct from Arthur Sweetnam's field near the red metal footbridge. Rover, our collie dog, always followed behind, to protect us, as we neared the viaduct – our railway holy grail.

When I was young and in the same train as Joe St Leger, a family friend since the mid-1950s and famous railway photographer, I noticed that he wrote everything down in a brown diary. From that very moment I wrote everything on Browne & Nolan jotters; without my records I would certainly not have remembered all the events, details and people that have proven essential in the writing of this book.

The nice thing about having an interest in Irish railways is that I have met great and loyal friends, like the aforementioned Joe St Leger, Colm Creedon, Walter MacGrath, Ray Good and David Bate. Walter MacGrath and Tony Price used to give outstanding Irish Railway record society slide and film shows on the Cork and Kerry railways during the 1970s and 1980s. I learned a lot from them. In time, I had to take over from Walter MacGrath (the former Irish Railway Society lecturer) when he unfortunately died. It is now nearly time for someone to take over from me.

Railways everywhere are valid sources of social history. Trams and light rail will hopefully make a comeback in the greater Cork area, like the closed Harcourt Street line in Dublin that reopened in a revised form as part of the Luas. A light rail system westwards to Ballincollig and Ovens would make a good start. The population of Cork – and all of Ireland – deserve better.

My father, mother and I alight from carriage 986. Colm Creedon informed me much later that he was lucky to get this shot as the lady nearby kept getting in the way with her sway.

BIRTH

THE BIRTH OF THE WEST CORK RAILWAY

Beginnings from Ballinhassig to Bandon

Railway and transport history is in the air these days, particularly in Cork, with over 1,000 people turning up for my last two railway shows in Bandon and Ballinhassig – the places where it all started in the 1840s. Railway mania first took hold when Ireland was still part of the United Kingdom when the strong business town of Bandon sought and got a railway bill presented to parliament in London in 1845. Queen Victoria declared the 1845 sessions open on 4 February at 10 a.m. and the Cork and Bandon railway bill became law on 21 July.

The proposed route for this railway line was planned by Edmund Leahy, who was appointed resident engineer under consultant engineer Charles Vignoles. The route started in Cork city from Albert Quay – the Black Ash – Pouladuff – the Doughcloyne Delta – Chetwynd – Castlewhite – Ballyma – Waterfall – Goggins Hill – Ballinhassig – Killeady – Crossbarry-Garryhankard – Brinny valley – Innishannon – Bandon. Despite severe curves and gradients along this route, it was accepted and on 16 September 1845 the Earl of Bandon turned the sod at Kilpatrick amidst the cheers of peasant, peer and peelers. Following this, work started immediately on the Bandon-to-Ballinhassig section.

This intense engineering project involved constructing the Kilpatrick Viaduct and tunnel (Ireland's first), a three-arched stone viaduct at the Halfway, Ballinhassig, as well

Albert Quay terminus.

as cuttings, culverts, massive embankments, track and ballast laying, etc.

Countless mules, jennets, horses and donkeys were used for the transportation of earth, stone, timber, rails, sleepers and of course people and their daily needs. This was like a Klondike with a capital K – in fact a Special K. Temporary stables, stores, forges, tents, grocery wagons, including supplies of butter from North Cork, were needed. Other services necessary for the undertaking of this project included medical tents, toilets and veterinary wagons. The wages wagons that were used to transport wages were made of iron and were protected by the police due to the danger of gangs of bandits attacking them on paydays, which happened several times.

De Courcey's public house (now Alan Barry's Ramble Inn, Halfway) did a roaring trade during the construction. The Irish navvies drank Beamish stout, Lane's Cork porter and Irish whiskey. The English and Welsh labourers drank mostly ginger beer. Horse drawn hackneys, official and unofficial, carried people to and from the various workplaces and also to pubs and churches. A wooden cross was also erected, which became a place of prayer as the nearest church was two miles distant. It was near Milepost 11, which was often referred to as 'the Ledge' – there was a saying that you 'took the pledge at the Ledge'.

This source of work helped to partly counter the effects of the famine, which cost millions of Irish lives. There were very few ships to take the stricken population, if they could afford it, to the safe shores of America, Spain, Italy or France. A far greater figure death figure would not surprise me even though the official figures are 1 million dead, with 2.5 million people forced to emigrate. Hundreds of workers, including children and animals, sweated and toiled on this difficult ten-and-a-half-mile stretch.

ABOVE: An 1850 drawing of some of the work force [L to R]: Robin Bickerstaffe, Cyril Harbatch and Cecil Thruslove drinking at De Courcey's Bar, Halfway, Ballinhassig, after a hard day's work on the three-arched stone viaduct at Ballinhassig.

Among those workers were former Allihies, Urhan and Eyeries miners who'd left their former roles due to a downturn in copper mining activity in the Beara area. They worked on the Kilpatrick tunnel project, along with the West Carbery miners. When both the Kilpatrick and Ballinhassig tunnels were completed their next destination was Butte Montana in America where the descendants of the O'Sullivan's, O'Shea's and Harrington's still live.

ABOVE: Albert Quay station, built 1851.

RIGHT: The earl of Bandon cutting the first sod.

BELOW: Chetwynd viaduct 1851.

After track and ballast had been laid, the next stage was the introduction of locomotives and rolling stock. Locomotive engines in kit form, rails, sleepers and tools were shipped from the midlands of Britain to Colliers Quay, Kilmacsimon. Horses then drew these to the various sites between the construction sites of Bandon and Ballinhassig.

As the Ballinhassig-to-Bandon line neared completion, the resident engineer, Edmund Leahy, was in conflict with his employers over some dubious decision making, like his advice to use light rails on longitudinal sleepers, which was very dangerous. Heavy rails on sleepers were the safest option at the time, and continue to be today. As a result of these tensions and some other unwise decisions, Edmund Leahy 'received his coat and billycan' (a railway term meaning he was sacked).

His replacement, the chief engineer Charles Nixon, drove the first Ballinhassig-to-Bandon locomotive on 30 June 1849. At 12.30 p.m. the crowds cheered as engine No. 2 *Sighe Gaoithe* (Whirlwind) steamed out of the flower-and-laurel bedecked station. The official mascot for the day, an Irish wolfhound, barked from the platform as the train left – and no wonder he did because he wasn't allowed to board the train! On arrival in Bandon, the guests were fed and watered at the Devonshire Arms Hotel.

The first journey had been a complete success; now it was time for phase two.

BELOW: Bandon station opened 1849.

LOWER RIGHT: Cork Bandon and South Coast Crest.

BELOW: West Cork train at Albert Quay in the early twentieth century.

Albert Quay to Ballinhassig

Fox Henderson & Company of the London Crystal Palace fame won the Cork and Bandon railway contract, which involved the construction of the railway line from Albert Quay in Cork city to Ballinhassig and included dozens of culverts, bridges and a tunnel at Goggins Hill. During 1850, work went ahead at a vicious pace, the work trains steaming at full pace between Albert Quay terminus and Ballinhassig. As well as that, two four-horse-drawn coach omnibus vehicles plied their trade from the Imperial Hotel Cork to the Ballinhassig 'bus road'.

This speedy pace was continued up to 31 July 1851 when an issue arose that dismayed the contractors: they received no payment from the railway company due to low cash flow.

Matters quickly came to a head at Milepost 3 near Doughcloyne and also at Milepost 4 at Chetwynd when fencing, rails, ballast and sleepers were removed and damaged by agents of Fox Henderson & Company. The constabulary arrived from Cork city and Ballinhassig and dozens of arrests were made in the darkness of night. The saboteurs had to be chained to an old oak tree to sober them.

It was alleged by Fox Henderson that the substantial sum of €30,000 was owed by the Cork Bandon Railway Company. Just like the modern-day political tribunals, the subsequent court case following the attack on the railway proved to be a financial bonanza for the lawyers involved. The crown briefed both the defence and the prosecution for the case, which opened on 12 August 1851.

Cork and Bandon Locomotives 1849–1899.

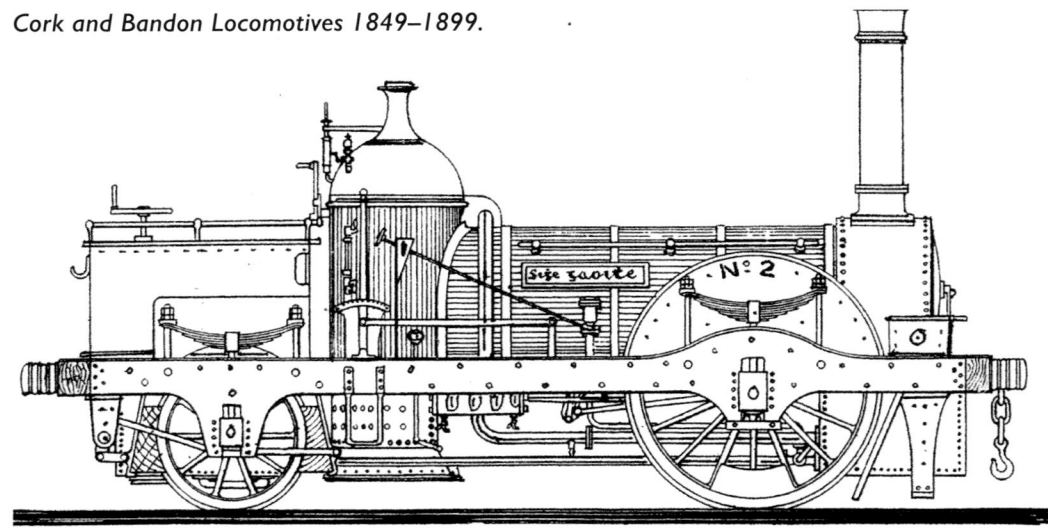

Sighe Gaoithe.

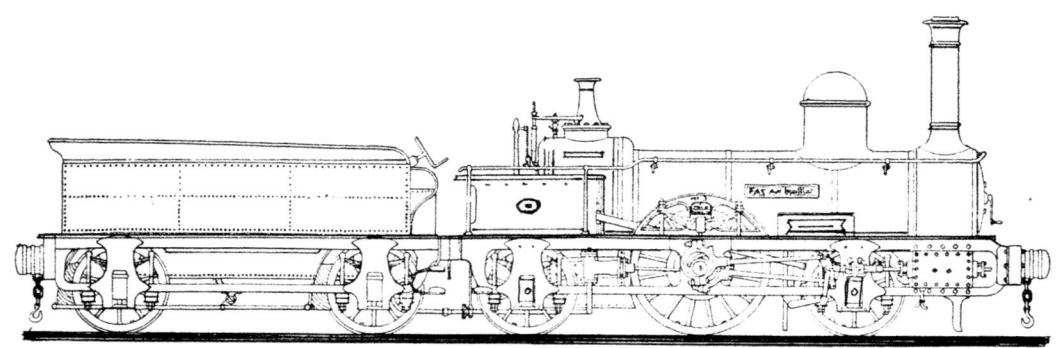

Fag an Beallach.

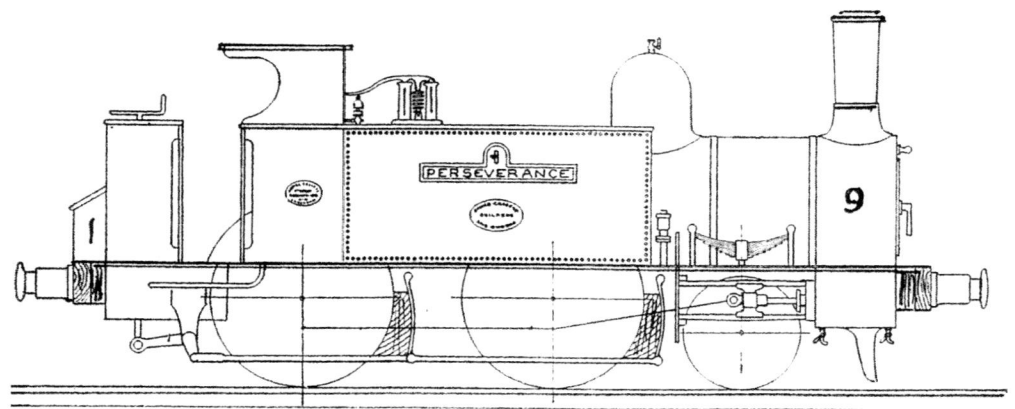

Engine No. 1.

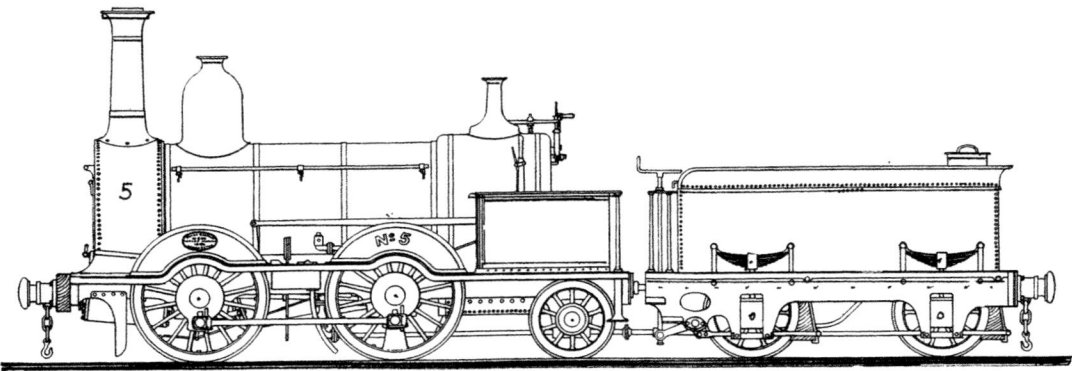

Engine No. 5.

GSR Engine 274 resting at Drimoleague.

The celebrated lawyer and parliamentarian Isaac Butt, of the Home Rule movement, was employed by the defendants Fox Henderson. The charges included damage to the viaduct at Chetwynd, using dynamite to blast rock at Ballymah and the dumping of earth and stone at Milepost 3 at Doughcloyne. The court erupted in laughter when a railway agent J. O'Leary remarked, 'We marched in song from Goggins Hill together and divil a sight we saw of our enemy'.

Judge Baron Pigott, who was recognised for strict impartiality and conscientiousness in his arbitration, read out the verdict of the jury. A very surprising not guilty verdict was declared in favour of Fox Henderson. The Bandon Railway Company did not appeal because British and Irish legal advice was proving to be far too expensive.

With the sum presumably paid, progress at Goggins Hill tunnel proceeded at a very rapid pace following the trial. Meanwhile, at the other end of the line, the Brunel-inspired Albert Quay station was finished with fine East Cork stone to the engineering standard of Brunel, who was one of the most respected and prolific figures in engineering history, and who gave solid advice to his former pupil Charles Nixon on the Cork-and-Bandon line.

5 Engine No.2 as rebuilt.

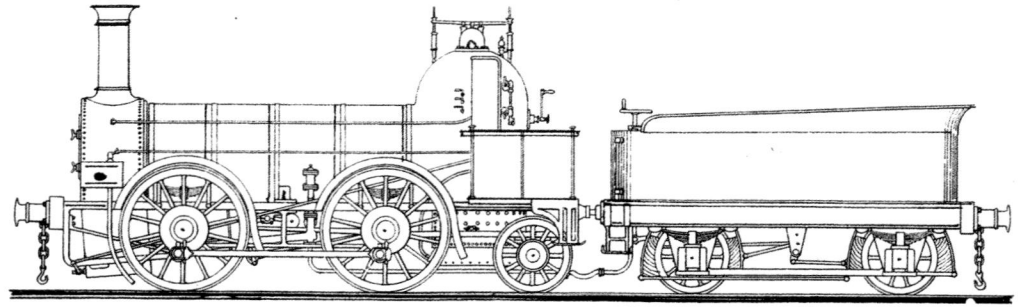

Engine 3 No 1.*

The first Cork-to-Bandon train departed platform one at 12 p.m. on 6 December 1851 with 290 passengers aboard, including the directors, the public and, yes, you have guessed it – the politicians. It was here in 1851 that the Engineer Charles Nixon declared this great railway to Bandon will last a thousand years to serve the people of this great country.

This successful enterprise was the direct result of brave far-sighted and civic-minded people who brought Cork and Munster directly into the twentieth century with a then-state-of-the-art, inter-connecting railway system. In fact, this enterprise proved so popular that Bandon became a Mecca for horse-drawn coaches, eager to offer train travellers the option of travelling on to places further west, like Skibbereen, Bantry, Glengarriff.

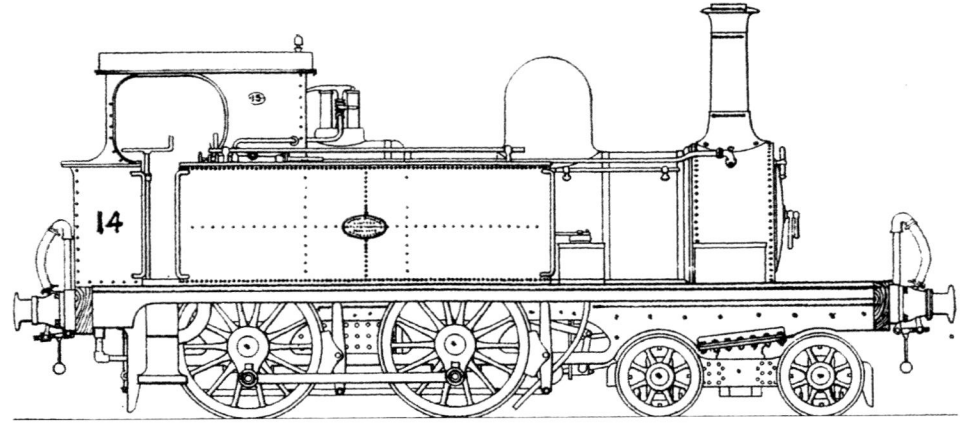

Engine No. 14.*

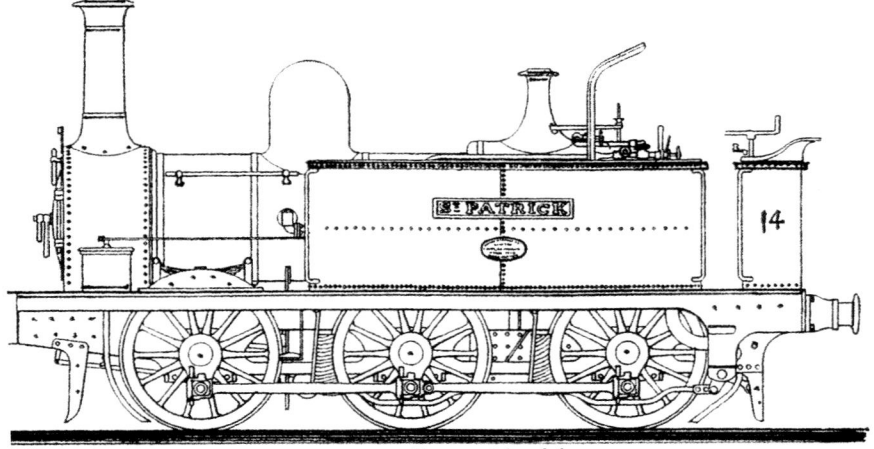

Engine No. 14.

*Where there are duplicate numbers, they were replacement engines.

EXTENSIONS

The first railway extension to serve Kinsale officially linked on 27 June 1863 and it branched off from the Cork-to-Bandon railway at Kinsale junction at Crossbarry where it wound its way in a semi-circular fashion to Ballymartle and Farrangalway, before travelling over a steel and stone viaduct and entering Kinsale via a rocky ridge.

Bandon to Dunmanway was the next extension to be traversed on 12 June 1866 with locomotives aptly named 'Perseverance' and 'Patience'. A further extension from Dunmanway to Skibbereen soon followed.

The Bandon Railway Company was not operating in isolation at the time, however, as the Cork and Macroom direct railway was incorporated in 1861 and its chairman was Sir John Arnott and the engineer for the railway was the noted architect Sir John Benson. It opened on 12 May 1866, utilising the Cork Bandon Railway terminus at Albert Quay.

The company was charged the sum of £2,000 for the privilege, an enormous sum at the time. This branch left the main Bandon line at Milepost 1 before turning right, heading for its first stop at Ballincollig. This station was extremely lucrative, bearing in mind its proximity to the Ballincollig army barracks.

Further stations on the line included Bishopstown, Ballincollig, Killumney, Kilcrea, Crookstown Road Dooniskey and Macroom. Trains operated from Albert Quay from 1866 to 1879 and again from 1925. (From 1879 to 1925 trains ran from Capwell Railway Station because of strained relations with the Cork and Bandon directors.)

BELOW: Irish troops with their Stahlhelm-style helmets at Grange level crossing near Ballincollig during the emergency years. The locomotive is possibly a Bandon tank 4-6-0. Note the metal lattice signal later moved in the 1950s to the Bandon section by the PWD red flat wagon.

The Lee Valley Viaduct with an Ivatt box top locomotive hauled freight train complete with drovers' green carriage.

Ballincollig station handled heavy goods and military traffic because of the army barracks nearby. Arms, ammunition, horses, food, linen, tools, newspapers and various other supplies came by rail. Even the rail tickets to Ballincollig had nicknames: the 'Brit pass' was a military ticket, while the 'Paddy pass' was a ticket issued to locals.

The first Ilen Valley Company train from Dunmanway entered the town of Skibbereen on 23 July 1877, although amidst a slightly lesser welcome than usual. The next extension was to beautiful Bantry Bay, which was reached on America Day, 4 July 1881 with a special train from Drimoleague junction.

Clonakilty sought and got its railway from Milepost 23 Gaggin on 28 August 1886.

West Cork's first narrow-gauge (lightly laid) railway was established with the construction of the Schull and Skibbereen railway on 9 September 1886. This gauge was three feet as opposed to the 5'3" broad-gauge.

The Cork and Bandon Railway officially changed its name to the Cork Bandon & South Coast railway in 1888, around which time the Bantry line was itself extended to Bantry pier.

Ballinascarthy became the system's fifth junction (after Macroom, Kinsale, Clonakilty and Drimoleague junctions), when it was linked to Timoleague on 23 December 1890. A Santa Christmas special train ran from Ballinascarthy to Timoleague and wooden toys were given to the children to mark the occasion.

Another development was the opening of the Ballinhassig aerial ropeway between Ballinhassig station and Ballinphellig, which was a distance of three and a half miles. This aerial engineering marvel had large red-and-green buckets travelling by high tensile wire over hills, bogs, streams, rivers and dusty roads. The tough trestles were twenty-six feet high with a four-feet-wide way, allowing speedy two-way bucket traffic carrying bricks and Welsh coal. This was all done to service the local brickworks, which the parish priest opened in the summer of 1901 amidst great local scenes. John Sisk was the manager of the Cork Brick Manufacturing Company. Each bucket carried a half-ton of red brick, which was the equal of the finest Lancashire rose bricks. The average daily load was 160 tons of fine quality brick for the Irish market.

Six workers toiled on the red bricks at the railway station while twenty local labourers worked the brickwork's extremely hot furnace. When the foreman was not around the lads did foxers (unofficial work), making shovels, hammers and small gates, etc. The whole operation was built by the famous Roe & Bedlington Company London. Rail conveyance of bricks to Cork city and beyond was far more efficient than the previous method of road transport when the twenty-ton brown St Patrick steam road engine with its blue trailers tore the dusty roads to pieces. When the whole operation had to close in 1913, there was a serious plan to make a tourist cable car venture out of the ropeway. Alas, the First World War put paid to the plan.

In 1912, further advances in the railway saw West Cork connected to Dublin, Belfast and Derry. Another great industrial venture started at the Marina in Cork city during 1917 when Ballinascarthy's proud son Henry Ford created his huge motor assembly works there, complete with a direct rail link to Albert Quay station, Cork. Trains sent Ford products to Dublin, Belfast, Derry, Sligo, Galway, etc., in record time. Ford car dealers, etc., found that this new form of transit brought costs lower and lower so that on average people had more money in their pockets, thus creating more jobs.

It was not all plain sailing, though. For example, the Macroom branch suffered during the Civil War of 1922–23. The anti-Treaty forces committed serious railway vandalism at places like Curaheen, Ballincollig, Killumney, Dooniskey and Macroom level crossing, but thankfully there was no loss of life.

The final meeting of the once-proud directors of the Cork & Macroom Direct Railway took place on 28 February 1925 before handing the railway over to the GSR on government orders.

Still, these extensions continued well into the twentieth century.

1925 GSR enamel sign, which is now preserved in Bavaria. The other two signs are in West Cork and the United Kingdom.

In this photograph we see a Ford F. 100 hightailboard lorry, a Leyland route No. 8 bus as well as a freshly painted No. 90 loco at Brian Boru Road and Rail Bridge.

Coach no. 1086 in Ballinhassig.

GSR & CIÉ Era

The Great Southern Railways (GSR) was formed in November 1924 with the forced amalgamation of The Great Southern and Western Railway, The Midland Great Western Railway, The Dublin and South Eastern Railway, the Cork Bandon and South Cast Railway, the Cork and Macroom Railway and various other smaller railway companies.

Shortly afterwards, in 1931, the Kinsale railway was closed as the GSR was very much strapped for cash. It had never been a great success as the Kinsale station was geographically challenged, being located on a steep hill outside the town. When the British left in 1922 the amount of passenger traffic on the route had been reduced by thirty-eight per cent and the situation continued to deteriorate so by 1931 the GSR had no economic choice but to close the station. The Kinsale line was not the only line to suffer during this period. Because of cheaper bus fares, the recession and emigration to the USA, the passenger trains on the Cork-to-Macroom line were also terminated in 1935.

Another closure occurred during 1953 on the Cork-to-Macroom Railway because the ESB was constructing two hydro-electric power stations on the river Lee which would flood the railway from Doonisky to Macroom Milepost 22. The Doonisky enamel name board was preserved by the late Joe St Leger, who also saved many other Cork railway signs.

Córas Iompair Éireann (CIÉ) was established in January 1945 and fully nationalised on 1 June 1950, retaining its very capable chairman T. C. Courtney who had earlier worked for the Cork Bandon & South Coast Railway before finally joining Henry Ford & Son at the Marina plant in Cork city. This able CEO understood main line and branch line operations very well and he did not tolerate any political interference. He ran this great transport empire like clockwork despite CIÉ being cash-starved. He commanded loyalty and respect from the vast majority of the large workforce under his command. As a railway manger told me, 'The Last Emperor' of the transport empire was T. C. Courtney, as he was an honest and loyal railwayman.

He proposed a solution to guarantee the future of the West Cork mainline during the 1950s. Number one was to replace the costly steam locomotives with diesel engines. Number two was to close the Baltimore, Clonakilty and Courtmacsherry branch lines to save on costs. In 1953, he warned that, unless pruned, these branch lines would close the entire mainline railway from Cork to Bantry. However, the powers that be of Clonakilty, Skibbereen and Courtmacsherry rose up against this very sensible proposal. Unfortunately, he backed down because of this pressure, which he later admitted was a

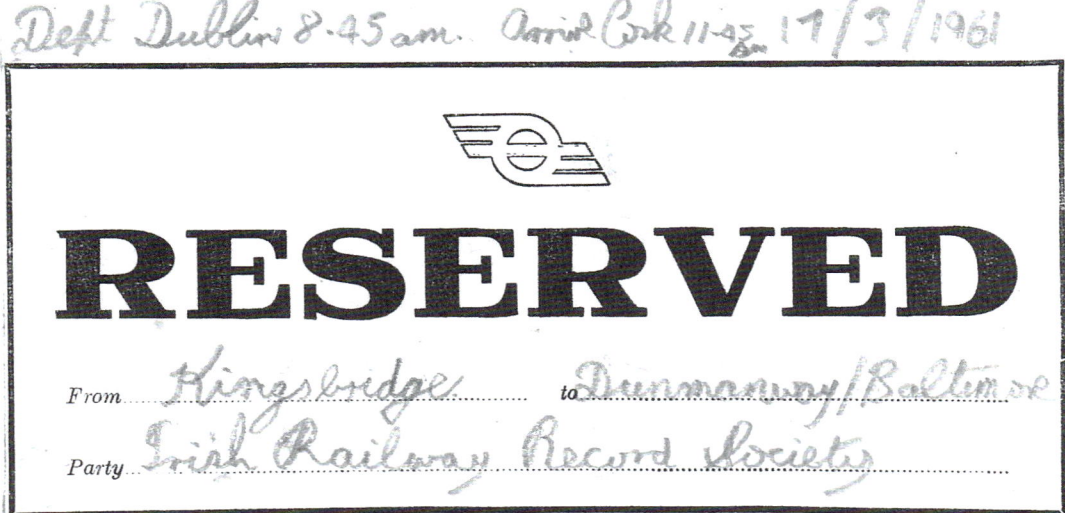

CIE window sticker from 1961.

mistake. When Cork Airport was in its planning stage, T. C. Courtney also proposed a future rail link to the new airport, as well as future stops at Togher, Doughcloyne and Chetwynd. He was way ahead of his time and he also suggested a rail link from Shannon Airport to the Ennis–Limerick line.

T. C. Courtney, had another major dilemma during the 1950s as the Macroom sleepers were rotting and most of the rails were rusting and he wrote to the railway and tram historian, Walter MacGrath, stating that while in the far distant future he could see a need for a suburban line as far as Ballincollig, there was no money in the kitty. The ESB solved his dilemma by announcing the construction of two hydro-electric power stations on the river Lee. During this time there were Howard's grain trains, trains from Ford's with tractors and cars as well as livestock specials.

Córas Iompair Éireann was green in every way with a capital G during Courtney's tenure, with two-tone livery on their trains, trucks, buses, buildings and ships. CIÉ was a corporate masterclass with the added addition of their logo masterpiece: the winged wheel. Many designers believe that the CIÉ logo was on the same level as the world's best commercial logos, such as Ford, London Transport, Opel, Esso, Shell, RCA-Victor, Coca Cola, Guinness, British Rail, Aer Lingus, Mercedes Benz, VW, BMW, Cadburys, the RAF roundel and the German iron cross.

Despite his reservations, T. C. Courtney promised both Walter McGrath and Colm Creedon that he would revive the Cork-to-Courtmacsherry seaside excursion trains, a progressive promise that he kept. He went even further during 1954 when he awarded Bantry with a dark green AEC railcar set, number 2641. Silver 'C' Class Locomotives later worked the mainline and branches of the West Cork system. Loco trials with the A and E class locos occurred also around this time

When Courtney resigned as chairman of CIÉ on November 1958, it was the end of an era.

'NIGHT OWLS' ANTI-SOCIAL HOURS

The work began at dawn at the Mallow Sugar Beet factory, with both silver metal and grey timber wagons coming from various places like Timoleague, Ballinascarthy and Clonakilty Junction. The CIÉ railway men who operated the beet operation were nicknamed 'the night owls' as these West Cork-to-Mallow specials only ran at night and well in to the wee hours of the morning. The work itself lasted until after Christmas.

The first West Cork special (train) was followed at the Mallow Beet factory by a Midleton special with 200 tons of beet. The green CIÉ locomotive was uncoupled and replaced by a German-made sugar company locomotive and two methods were used in unloading the beet: a rusty grab crane unloaded five wagons per hour and the other method was a huge water gun which produced a very powerful jet of water, forcing the beet out of the wagon.

The beet then fell down in to an underground channel, which led into the factory itself where it was washed a second time, sliced, weighed and given a very hot dose of water. This removed all of the sugar, leaving only the pulp, which was then dried and mixed with molasses. This was used to feed the cattle of Ireland. Up to 2,000 tons of beet passed through the factory every day and every night.

West Cork–Mallow beet train.

FARMING AND THE RAILWAY

From the birth to the betrayal of this railway, farming was always a railway revenue constant. Trains transported horses, cattle, donkeys, pigs, poultry and dogs from farm to village, town and city. All the county fairs, such as Bantry, Timoleague, Kinsale, Macroom, etc., were expertly catered for by trains consisting of locos with open wagons, closed wagons, flat wagons, cattle wagons and horseboxes. Similarly, barley, wheat, corn and other forms of grain were carried from Bandon, Macroom, Crookstown, Shannonvale and all over Ireland by freight train. Sand and beet were also transported from Courtmacsherry and Timoleague to the rest of the country.

Flat wagons from the Henry Ford & Son plant at the Marina in Cork brought Fordson Major tractors, Thames Trader trucks and Anglia vans to both farmer and creamery alike, as well as the Consul 375 (Ford designed car from the 1960s).

New and second hand farm machinery and veterinary supplies came from far and

wide via Glanmire Road railway station. Wool from sheep, eggs from hens, milk from cows, butter, cheese, etc., were rapidly carried to customers all over the country. Businesses similar to Robert Macklin of Bandon imported second-hand tractors like the Massey Harris and the red David brown from the United Kingdom. Fuel for farm machinery was transported by train from Foynes Port to the Esso storage tanks at Bantry station.

Many second-hand railway sleepers found their way on to farms, having been retired and removed from the railway. Some worn-out rails and fittings from closed branches like Macroom, Schull, Skibbereen and Kinsale were also sold as scrap to the farmers for hay shed and outhouse buildings. Old former decommissioned Cork & Bandon timber box wagons always found a good home on farms.

Many a good Collie sheepdog or pup travelled on the train, complete with their own dog ticket, to far flung places like Down or Derry. Goulding's Fertiliser (a very famous fertiliser company) was loaded on to open wagons at Albert Quay for dispatch nationwide. It was a two-way system with goods and animals coming in and out of farms via the railway.

Further out the line towards Dunmanway – more pigs, sheep and cattle.

All the various railway companies from 1849 to 1961 transported people and animals to and from Cork's Albert Quay station. Racing pigeons were transported from Belfast, cattle from Cahirciveen and dogs to Donegal, etc. British army horses, marts, pets, also used the railway down through the years. Cats, dogs, horses, mules, donkeys, sheep, cattle, pigs, hens, ducks, geese, turkeys, peacocks, rabbits, pigeons, birds large and small arrived and departed from all stations. The train guard Dick Hudson often said to my mother that animals were quieter than humans on the train.

Inside the goods building in Albert Quay that housed the freight for Bantry, lived a pet crow called 'Joe the Crow' who used to venture out the line on the roof of the grey brake van to the 'Snotty Bridge' (present-day Kinsale roundabout) and would fly back later in the evening with his belly full. A murder of a woman took place on the small red iron bridge not far from the 'Snotty', during the very early 1960s. The same 'Snotty Bridge' was the scene of a brutal murder during the late 1950s. Unfortunately I could not find out how it got its name.

The circus also made use of the railway. The world famous Bertram Mills Circus came to Cork city in 1961 by rail and parked at Albert Quay. A very large complement of exotic animals, like llamas, pumas, monkeys, bears, tigers, chimpanzees, lions, leopards and last but not least the beautiful and humble elephant alighted from the circus train. The elephants, complete with baby, left Albert Quay station on foot with their friendly keepers to the nearby circus site, escorted by a black Ford mark II Consul garda squad car. The Ford vehicle train had to wait down at the Marina while this trek was taking place as the noise and smoke from the bright green Metro Vick C212 would drive the normally placid elephants amok, so never the twain did meet. Joe St Leger was on hand to cine film this thoroughly titillating trek and my father brought my brother and I to this Bertram Mills super show and every act was great.

LEFT: Cock of the walk at Crookstown.

Paddy, the horse, shunts a wagon of grain at Shannonvale Mills, near Clonakilty, 1961.

LEFT: Vince Milne gave Paddy a good home after retirement. To quote Vince 'We saved him from the dog food factory and looked after him in the only way we could'.

ABOVE: Hughie Harrington's pony and dray.

RIGHT: 'Joe the Crow' perched on the Snotty Bridge in 1960, having hitched a ride on the roof of the brake van from Albert Quay.

BELOW: Cattle, sheep and pig train at Togher.

1888 The Prince of Wales early tourism route promotion stand.

TOURISM

Tourism started in a big way in 1858 when Edward, prince of Wales, decided to travel the West Cork rails. Upon his arrival in Cork, he travelled in a royal blue state horse drawn coach to Albert Quay Railway station. Laurels had been brought up from Innishannon to decorate the station concourse. The band struck up marching music while the guests were provided with tea, cake, gin and fresh Caviar before their departure on the rail.

Meanwhile out in Togher the people who laboured hard in their market gardens considered themselves very lucky to get some semblance of a small dinner and water. The local police chiefs and military were furious over all the hype because they had to draft scores of men to protect the route from Albert Quay to Bandon. Two policemen were attacked at Doughcloyne railway bridge (Milepost 3) by a band of locals seeking arms and ammunition of any kind. Another gang with cudgels were beaten off with gunfire at the Abbey Bridge outside Waterfall. 'Give them the lead' shouted the sergeant who was known as Buckshot Bill.

Extra armed military travelled on the purple lake coloured royal railway carriage which reached Bandon without further incident. For some reason Prince Edward called Bandon 'Bandonia'. He then travelled to Killarney via Clonakilty, Dunmanway, Bantry and Glengarriff by a horse drawn omnibus in crimson livery. The Cork Bandon Railway Company heavily promoted the 'Prince of Wales route' between Cork and Killarney. In terms of tourism, this unprecedented tour presented West Cork as a must-see destination for the British aristocracy and its wealthy followers. As a result, it generated revenue well into the twentieth century.

Cork tourism got a great boost during the 1880s when the Bantry Bay Steamship Company ran two goods and passenger ships, the *Lady Elsie* and *Princess Beara*, from Bantry pier to Glengarriff and Castletownberehaven. The steam-ship company's original building still stands on Glengarriff Pier and is well worth a visit. Ireland's spectacular scenery was viewed from these fine and robust ships up until the 1940s. The *Lady Elsie* was always escorted into the harbour by the local dolphin, named Florrie Batt by the fishermen, although for some reason only known to her she never escorted the *Princess Beara* into the harbour.

West Cork tourism got another boost in 1912 when the Great Southern and Western Railways were connected to Albert Quay. From that point on, if you left busy Belfast after your early morning sausage and rasher breakfast you could top up your empty food tank in Vickery's hotel Bantry later that night, washed down by a pint of Beamish porter.

Because of all the wars – First World War, War of Independence, the Civil War and the Second World War – tourism took a steep dive until the very late 1950s. The original term 'The Wild Atlantic Way' was coined in the Eccles Hotel Glengarriff during the 1890s by an English tourist and was revived again recently. Sometimes what has been done before and hidden can later be rebranded. Unfortunately the unexpected closure and rapid destruction of the Bantry line diverted the rail borne tourists to Killarney, Dublin and Galway. *Wahre Holle* (pure hell).

'The Golden Triangle' which takes in Bantry, Castletownbere, Allihies, Eyeries, Ardgroom, Kenmare, Sneem, Valentia Island and Killarney was, and still is, Ireland's most scenic region.

MAP SHOWING THE PRINCE OF

WALES' ROUTE TO KILLARNEY.

Timecheck.

RAILWAY BARS

Walter McGrath, a noted railway historian, once said that all of the Irish railway bars were 'dens of iniquity'. I disagree, as most were social centres, so one would never quite know who would call in or drop out. Most stations on the West Cork rail system had public houses nearby. Waterfall was near Connie O'Shea's bar. Crossbarry had William Barrett's bar and lounge. Upton was served by Cronin's Railway Bar – now run by Imelda Twohig – whilst in Bandon, Kitty Crowley's was connected to the station yard. Hard-working men slated their thirst with Beamish and Murphy's whilst the *garsún*s had Little Norah lemonade in these institutions. To name all these wonderful community pubs along the railway would take too long.

These bars often saw an interesting – and sometimes famous – visitor. For example, when making the film *Moby Dick* during the 1950s in Youghal, Gregory Peck boarded the railcar from Albert Quay to Crossbarry Station. His next stop was O'Callaghan's Crosspound Bar between Barrett's Bar in Killeady and Hickey's Bar in Aherla. The party went on well into the wee hours of the morning. Robert Mitchum also drank here, having finished the film *Ryan's Daughter* in the Dingle Peninsula during 1960s.

RAILWAY APPAREL

From the birth of this railway to the very bitter end all railway men took extreme pride in how they looked in their uniforms. Pants were creased, badges gleaming, buttons shining and boots polished. They walked tall, straight and looked the world right in the eye. The Córas Iompair Éireann staff looked very elegant in their black clothes and pointed leather shoes. Up to the mid-1960s one important item was the rather heavy waistcoat complete with flying snail brass buttons. The trouser belt was a very heavy type that replaced the earlier over-elaborate braces. The earlier shirts had removable collars, which were Cork made. The rain-proof oilskins were a disaster due to internal perspiring and their restrictive and tight fit which did not suit manual work or walking the line.

FURY ON THE RAILS

Billy Fury was one of the biggest music stars of the 1950s and 1960s. Connie Lucey, the then bishop of Cork and Ross, barked from his pulpit most high about the gyrations and loud music of Liverpool's and England's top 'rock' singer.

Billy had been banned in Dublin in October 1959 because of his very wild stage act. For his January 1961 rock'n'roll Cork City Hall gig he had to wear a suit and tie instead of his usual leather jacket and navy levis and turned up jeans. He wrote all his own songs and was the first singer to sing blues and R&B on a Cork stage. The old Merseyside docklands song which pleased the Cork rockers that night was 'Boozing'. Kevin Beale, the famous Douglas tunesmith said, 'everybody was singing that song for weeks afterwards'.

After his Cork performance, Billy Fury went to Bantry by train the morning after the concert to see the stunning scenery, the wildlife and the birds, the feathered kind that is. During 1960, The Beatles failed a test as a backing band for Billy Fury and John Lennon was photographed getting Billy's autograph.

Billy was not the only musician availing of the railway. Bands such as the Butter Exchange and the Barrack Street Band complete with their huge array of instruments often travelled the mainline train to Bantry and beyond to parades, races and regattas. When Guard Dinny Keane was unloading the big base drums he gave a few gentle taps someone shouted 'Docha Dinny boy' in fine Cork slang. The passengers did not have to be in the height of porter to sing during the hour-long train journey from Cork to Bantry. Card playing, knitting and of course reading were also popular.

I used to look out the large window and absorb the magnificent views and sometimes a transistor radio would just about blare out some Elvis rock song which was great, but more often than not, you were subjected to Brendan O'Dowda the Irish tenor, Bridie Gallagher and Joe Lynch; nonetheless the instrumental guitars of the Shadows saved the day. 'Apache' sounds better today than it sounded back in the 1960s, viva Hank Marvin.

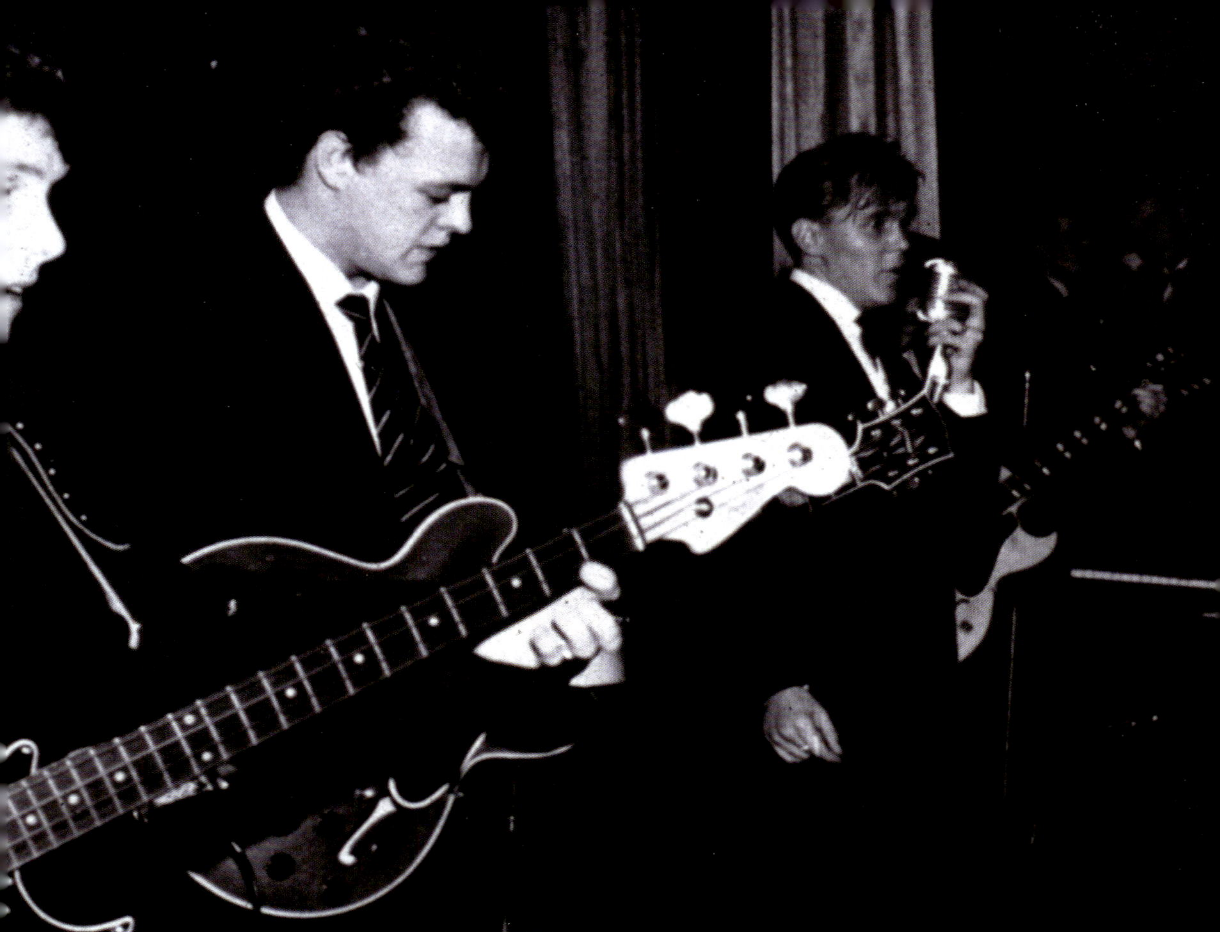

January 1961 at Cork City Hall – a rock, country, blues and gospel night with Billy Fury who railed it to Bantry the following morning. He had Rheumatic fever as a child, which weakened his heart, and he died of a heart attack in 1983 (aged forty-two).

RIGHT: New Musical Express *from the bookstall 1961 featuring Billy Fury on the front page. Also on sale were the* Southern Star, Cork Examiner, Irish Press *and* Irish Times *newspapers. The sweet tooth was well served by Cadbury's chocolates, Urney chocolate, Rolos, Kit Kat and Smarties. Tanora and Little Norah were the soft drinks sold here.* The Dandy, Topper, Beano *and girls comics were well placed at the stall. Books about Michael Collins, Tom Barry, Winston Churchill, James Dean, Erwin Rommel (the Desert Fox) Christy Ring, Elvis Presley (The King) and a considerable amount of novels were also all available.*

NME CONCERT IN WORDS AND PICS

EDEN KANE ★ CONNIE FRANCIS ★ TONY NEWLEY

Plus fullest pop news service

THE WEDDING

ANITA BRYANT · MALCOLM VAUGHAN

TEARS ON MY PILLOW

THE McGUIRE SISTERS

new MUSICAL EXPRESS

WORLD'S LARGEST SALE — ABC certified figures

THE MOUNTAIN'S HIGH

DICK and DEEDEE

LONDON

TWO IN THE TOP TWENTY

GOING LIKE FURY!

BILLY does it AGAIN!

His Latest Smash Hit

JEALOUSY

on Decca 45-F 11384

No. 15 IN ITS 2nd WEEK

A PERSONAL MESSAGE FROM BILLY:

To all my loyal fans and everyone at Decca,
'MY SINCERE & GRATEFUL THANKS'

P.S. Also for your wonderful reception at Sunday's "NME" Concert at Wembley—"WOW", it was one of my biggest thrills—I shall always treasure the memory.

HALFWAY TO PARADISE

on Decca 45-F 11349

Still No. 14 after 18 weeks in the charts!

and now —

THE GREAT NEW BEST SELLING LP

"HALFWAY TO PARADISE"

on Decca ACL 1083

WITH 12 SENSATIONAL SIZZLING TITLES YOU'LL LOVE

• Personal Manager LARRY PARNES says:
• "CONGRATULATIONS, BILLY BOY"

• Sole Booking Representative TITO BURNS says:
• "WELL DONE, BILLY"

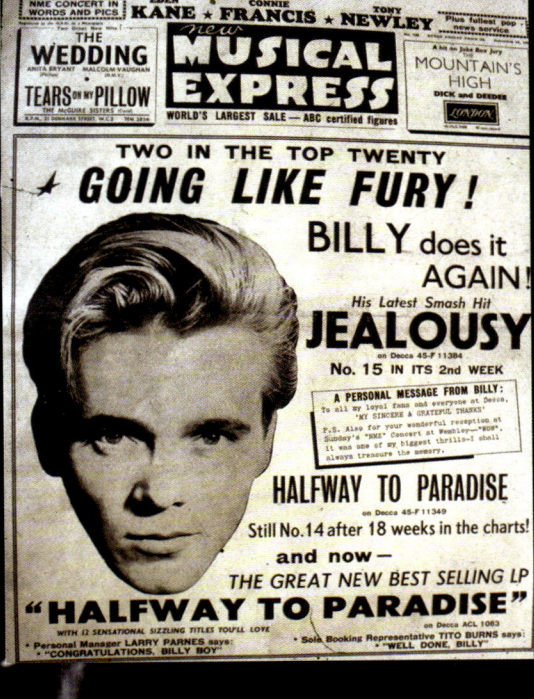

Chris with steam engine in Cork.

INCIDENTS

With the exception of the very last sad day I only ever once saw a garda arrest someone – and he happened to be dressed as a monk. He was well oiled (drunk) and went without a struggle, someone called him 'a caffler' (impertinent). Questions were asked about how did he get on the train in the first place.

Once a trainload of Ford Consuls and Thames vans were left overnight at Albert Quay because of a C class diesel engine failure. There were red faces in the morning when it was discovered that every single hub-cap had vanished into thin air. From then on all the hub-caps were locked in the vehicles.

Another time Henry O'Brien from Togher village walked out the line to Chetwynd Viaduct to view birds' nests on the trees, hugging the Viaduct on the Castlewhite side. Like rock star Billy Fury who had passed by a few weeks earlier, Henry studied ornithology and foolishly reached over the side of the 100-foot bridge. Fortunately he did not fall to his death but landed on the branch of a seventy-foot tree. He was spotted by Mr Twomey, a railway employee who alerted the Cork Fire Brigade. Henry learned his lesson and avoided heights after that near miss.

The top deck of the Viaduct was a very scary place to be when a full force gale was blowing and several adventurers had near misses down the years. When we were very young Noel 'Nollie' O'Meara, my brother Jerry and I would go to the viaduct and climb up and down the ladders from the decking to the pillar. Rover our Collie dog had more sense, and could not be induced to go on the 100-foot summit. We often threw down sods at the Teddy Boys on their trek out to the Chetwynd Viaduct and fortunately they could not get their hands on us as we were well above them. Their only response was a few well-aimed ballast stones, which came whizzing in our direction. We ran and jumped over a fence like horses over Becher's Brook.

TRAVELOGUE

All Aboard for Cork City to Beara

Our summer family outing to Bantry and Glengarriff when I was eight years old began at the Statue in Patrick Street where we boarded the two-tone green Leyland double Decker R678 bus to the Albert Quay rail and bus station. While the new bus station was being constructed at Parnell Place, Albert Quay station temporarily became the main bus station, which was a perfect Cork area transportation hub. One could hop on the train at Bandon and alight at Albert Quay and then take the red and white P class bus to (Crosser) Crosshaven. This option became so popular that people began to ask what the point was in wasting lots of money building a brand new bus terminal when Albert Quay station was perfect for both bus and rail operations. Even the taxi men believed that it was in a better location. I believe this was T. C. Courtney's only big mistake as head of Córas Iompair Éireann.

As my father bought our tickets in Albert Quay station, I cast my eye on the brown News Brothers newspaper and sweet shop. All the goodies were on display – Cadbury's dairy milk chocolate bars, Cadbury's Irish roses, Rolo sweets, Smarties, Fruitos, etc. Magazines like *Ireland's Own*, *The Fold*, *Elvis Monthly*, and *Woman's Way* were also on offer. A large selection of newspapers such as the *Irish Press*, the *Cork Examiner* were on view, as well as music magazines like the *New Musical Express*, which usually featured Elvis Presley, Billy Fury, the Shadows, Sam Cooke, Fats Domino, the Everly Brothers, Roy Orbison, Marty Wilde, Joe Brown and the Drifters. Nowadays it's simply called *NME* and, yes, Elvis is still the King of Rock.

You would know the city slicker by the clothes he wore, a very sharp look that featured narrow pants, narrow tie, the narrow collar and black Billy Fury boots with elastic instead of laces. In contrast, the 'Murty Dan' country look started with the Roches Stores' peaky cap, wide trousers and heavy brown hob nail boots, which made a lot of noise along the train platforms.

Paddy Hennessy, a station porter nicknamed 'the Beagle', was constantly patrolling, the station helping the public in any way he could. I had the unusual habit of picking up those beautiful designed Wills Woodbine and Players Navy cigarette boxes. I still have them, as well as the various tickets we used on our trips. Many commercial posters were pasted on the Albert Quay poster boards, such as Paddy Whiskey, Old Time Irish Marmalade, Cork Dry Gin, Guinness, Gold Flake cigarettes and various travel posters.

Gorgeous goods supplied by rail.

The big taxis parked outside the station were something else, like an American-built Ford Fairlane named after Fair Lane Cork City where Henry Ford's father stayed between his leaving of Ballinascarthy and his journey to the USA. Cork-built Consuls and mark II Zodiacs were also in attendance, as was O'Connor's beautiful black 1949 Dodge fluid drive V8. I saw my very first light blue Opel Record with German number plates at platform No. 2 with the owners wearing Alpine apparel.

The station flag with its winged wheel logo was fluttering as the time came to board the Bantry railcar. After a twelve-minute past midday departure, the train ran under the green and cream signal box before entering the Rocksavage tunnel, one of four on the system. On to the deep rock cutting under the High Street stone bridge before passing St Finbarr's Hospital on the left. We then passed various housing estates on the right. There was a serious gradient with a ratio of 1:100 for the next seven-and-a-half miles, which was no trouble for the Bandon tank locomotives or, for that matter, the AEC diesel railcars.

Milepost 1 was next. This was the former site of the Macroom junction, which branched off right up to the time of its closure during 1955. We passed over the Kin-

With 'Boots' Murphy at the controls, Chris Larkin observes the Iron Road to West Cork.

sale Road, under the Forge Hill Bridges and over the Lehenaghmore Bridge. We then passed through Milepost 2 at Togher, climbing steadily as we passed Milepost 3 at the Doughcloyne Delta – where I was born, as was my younger brother Jerry. Gaining speed, we passed Doughcloyne holy well, known as St Bartholomew's well during the famine times.

Passing Milepost 4, we headed for the Meagher's Lane Bridge; up next was the mighty Chetwynd Viaduct at Milepost 5 and then it was on to Twomey's Castlewhite level crossing, then passed some wonderful woods before entering Waterfall Station with its beautiful buff and white colour scheme.

After a nod to Signalman Maurice Tobin, we proceeded to the Abbey Bridge. We viewed lovely rolling hillside here because we were 410 feet high on this railway summit. Then we were going downhill all the way to Goggins Hill tunnel, where, from time to time, a lot of courting couples 'operated' inside this very wet and cold tunnel, hence its other name: 'the tunnel of love'.

The Ballinhassig station buildings were painted in the Cork colours of bright scarlet and white. This station had its own black and cream kissing (semi-circular) gate. Soon after leaving Ballinhassig, we went through astounding Swiss-like scenery past 'the Ledge' at Gortnaglough and down in to Killeady Milepost 13. Jim Lane's cream and Brunswick green Crossbarry station looked well. This was the former Kinsale junction, though the Kinsale line had petered out in the 1930s. Next stop was Upton station, painted in a light green and ivory white colour scheme. Across the road was The Railway Bar, an outstanding and necessary social institution.

We then proceeded to Rockford rock cutting, where silver had been found way back in the 1840s by the Allihies and Eyries miners. We then passed the beautiful Brinny Valley (Milepost 16) and entered Ireland's first-constructed railway tunnel at Kilpatrick. My brother Jerry didn't like travelling through this tunnel, but a bar of Cadbury's dairy milk chocolate usually appeased him, as did his green Hillman minx dinky and smaller yellow matchbox Zodiac Ford mark II model car.

We passed over the salmon-red Innishannon Viaduct, skirting the beautiful river Bandon nowadays sadly bereft and devoid of fish and pure clean water because of pollution. Passing over the 'China Wall' (a two-mile long wall between Innishannon and Bandon) section at Ballylangley, we sped inwards to Bandon's red-bricked station, which was complete with subway. Years later, Joe St Leger gave me the Bandon parcels office black-and-white enamel sign. It was here in this town that the infamous Major Percival put manners on a young Todd Andrews during the 1920s Bandon military barracks handover to the Irish. He kept Todd Andrews waiting for hours. Jack Lynch told me

A black and tan loco crossing the Phoenix Bridge, Bandon.

A CIE postcard.

in 1985 that this led Todd Andrews to a lifetime dislike of Bandon and West Cork in general. The Japanese Army put manners on Percival in Singapore during the Second World War – what comes round goes around.

Our journey continued on past the railway's original end point. We passed under Bandon's St Patrick's Catholic church tunnel and proceeded to Clonakilty junction, where all the buildings – bar the footbridge – were in mid blue and bright cream colours. A very young Ray Good and Donie O'Donovan (Clonakilty Junction workers) were on the island platform, complete with Donie's wonderful little dog Nip. Gaggin (the Irish name for the junction) had the longest footbridge on the Irish rail system. Donie later presented me with a Clonakilty excursion poster, which I still cherish.

Our train driver, nicknamed 'Boots', put the pedal to the metal aboard our AEC railcar No. 2640. We continued on the main line while the branch line curled off left for Ballinascarthy, Timoleague, Courtmacsherry and Clonakilty. Next up was Desert station, with its attractive buff and medium white painted corrugated buildings, including the booking office, toilets and waiting room. Dinty, a brown and white dog, always kept an eye on bikes thrown against the privet hedge near the GSR Desert enamel sign. A horse railway had run from here to a local mill in earlier days. The black pony was called 'Toe head' by its drover 'Bunty' Barrett.

Milepost 30 was next as we slowly entered the green and cream Ballineen railway station. I was always on the lookout for colourful artful posters, but I can't remember even one poster at this station; they were probably in the waiting area, safe from my prying eyes.

We then passed beautiful lush sylvan countryside with its vivacious variety of oak, beech and chestnut trees hugging our train as it merrily moved onward, honey bees abounding. Milleenig black bridge was crossed before we started slowing down at the double signal post. The rails on Milleenig Bridge exist to this very day.

We were then at the Castle of the Yellow River (Dunmanway to you and me). On this particular trip, round metal cans containing reels of film were taken off the train for the nearby Broadway Cinema. Marlon Brando's *On the Waterfront* was written on one can. Michael O'Donovan's historic Shamrock Bar in the middle of the town still boasts a vintage railway poster and other railway relics.

After Dunmanway, this beautiful countryside changed from Bandon-style, green-grassed land to rugged rock and gorgeous glen, complete with gorse, gannet and goose. We passed over silver streams and bog as we moved under the shadow of Gloundha glen defile. During 1991, bus driver Richard Lee and I walked this wondrous place

Innishannon viaduct.

and found hidden under ivy some old rusty relics of the Cork Bandon and South Coast Railway.

From Milepost 41–44 it was all downhill as far as the delightful Drimoleague Junction with Barney Deane and Jack O'Donovan working on the goods siding (freight tracks). A pretty coloured scheme of cream and green adorned this island-platformed railway station. My eyes gazed in wonder at an Aer Lingus poster, which showed a turbo prop Viscount Aircraft displaying the classic mid green and white shamrock logo. The branch line to Skibbereen and Baltimore went off to the left at this point, crossing the main Bandon-to-Bantry road that was complete with the ever-present pot holes. The red and green fuchsia consumed the pine railway fence as we continued on towards Bantry.

Aughaville, with its tin shed, was our next stop. Passenger trains stopped here by request only. A goods' store was notably missing at this station. Looking outside, I noticed that the level crossing had halted a black mark II Ford Zephyr it its tracks; it belonged to the well-known figure of Jerry the Yank.

Next stop was Durrus road railway station with its black and mint green colour scheme. No outdoor posters here but its log-cabin-style building was pure Americana. A red ESB Thames 400e van waited near the newly painted level crossing gates: truly a picture postcard scene.

It was now uphill through some stupendous scenery to Milepost 53 and from here it was downhill through magnificent mountainous territory as far as Milepost 56. The

A train leaving Bantry.

A train approaching Bantry.

original Bantry station abandoned track bed veered off to the right here, full of weeds and thistles. We passed under the Bantry aqueduct, curling around the green metal bridge and onto the black girder bridge over the Glengarriff Road, which sported a yellow Dunlop radial tyre sign.

At Milepost 58, the brakes were gently applied for a gentle stuttering stop at Bantry station. After a little over an hour, we had finally arrived!

No fancy pastel colours here as Brunswick green and cream were the chosen station colours. Heavenly Hungry Hill, Adrigole and the Caha mountains immediately drew the eye and charmed the happy traveller.

Outside the station, Ford black Consul taxis awaited patrons outside the station fence, which was made of solid Baltic pine. There were three very large American automobiles parked outside the railway station: a green Buick Roadmaster, a pink and ivory white 1957 Chevrolet and a large black 1946 Chrysler complete with a luggage rack on its roof.

As we wandered around the town, we got a lovely surprise from our parents – a Mackintosh bag of Scots Clan chocolate and toffee sweets. A Japanese-made Bandai-brand tin-plate toy Córas Iompair Éireann double Decker green bus looked very tempting to me inside a toy shop window, as did the Triang Hornby British railways LMS region toy train with its dark red livery encased in a beautiful box.

Train and Mike Cullinane's Ford consul car approaching Timoleague.

After a brief and happy visit to Murphy's pub and shop it was time to start the last stage of our journey. We board the cherry red and white P class single decker bus to Glengarriff. The CIÉ busmen wore different coloured caps (with white piping) from their railway brothers (with green piping). The bus terminated at the Black Cat Café, which is now an internet cafe. A CIÉ Leyland lorry passed on its way to Castletownbere Pier, followed by a red Ford V8 bringing freight from Bantry station heading to the Emerald Isle Mining Company's mine in beautiful Allihies.

After a final taxi ride through Castletownbere, we finally arrived at our destination: Allihies. John Terry O'Sullivan's red public house in Allihies had a wonderfully illustrated *Cork Weekly Examiner* on display, showing the Allihies copper mines and the miners working underground. Great local names such as Hanley, Kelley, O'Sullivan, O'Shea, Harrington, O'Dwyer, McCarthy, etc., still abound. As we wandered, we observed collie dogs herd sheep coming from misty mountains down to minute meadows.

Cork to Courtmacsherry

One of the other great West Cork railway journeys by the Larkin family and dog were the Cork to Ballinascarthy and Skeaf Sunday excursion trips – from where we would sometimes go on to Courtmacsherry. Unlike the ultra-modern Bantry section railcars, the coaches on the Courtmacsherry seaside trains were timber compartment coaches of some Victorian vintage, loved and adored by the English railway purists like John Langford, Michael Davies, the Casserley's, James Boyd and countless others who are too numerous to mention. The Irish purists like Alan and Leslie Hyland, Tony Price, Canon Skuse and Sean O'Brien enjoyed the vintage charm of these green coaches.

These Sunday morning excursions always started with mass at St Peter & Paul's church. Afterwards, with our spiritual tank full, we headed to the Long Valley public bar and lounge for a quick sandwich, before walking to Albert Quay rail and bus station. There was extra bustle with buses busily buzzing around the station's newly tarmacked yard. Passing the dark green and lemon ticket barriers, we would often see Joe St Leger with two cameras strapped around his neck. Joe always travelled on the very first coach, so he could take a photograph of those of us alighting from the train at Skeaf.

There was already a strong family link to this railway section. My grandfather, Jerry Cullinane, worked on the Ballinascarthy-to-Courtmacsherry section from the 1890s until an accident at Timoleague station during the 1930s finished his Cork Bandon & South Coast and great Southern Railway career. He even helped erect the GSR Skeaf

Pat Deasy of Clogagh awaits the Courtmacsherry special at Skeaf.

enamel sign at the halt in 1926 and also demolished the tiny timber shelter because of severe storm damage.

My uncle Danny Cullinane removed the Skeaf sign after the closure for preservation reasons. He lent it to the late Walter McGrath who in turn loaned it to the Irish Railway Record Society in Dublin and it is on permanent display in our headquarters near Heuston Station. As well as being nostalgic, it has both classic Gaelic and English lettering, an enamel design classic.

Sometimes we used Jack Hanlon's Hillman taxi to go to and from Ballinascarthy station. And on Sunday evenings, as we returned home from our excursion, my mother needed to flash a lamp three times from Skeaf Bridge to stop the 8.45 p.m. Cork-bound train.

This railway, being Ireland's most scenic, had one more thrill for us before reaching Albert Quay. Between Waterfall and Pouladuff, we were treated with a fourteen minute vista of the night lights of Cork city in all their cosmopolitan splendour.. The purr of the green Loco C 220, after her day's work, resonates with me forever.

When travelling this stretch of railway by daylight, you could view nine different mountain ranges covering five counties. These mountains were Knockmealdown (Waterford); Comeragh (Waterford); Galtee (Tipperary); Nagles (Tipperary) Ballyhoura (Limerick); Boggragh (Cork); Mushra (Cork); Derrynasaggart (Cork) and the Paps (Kerry).

A silver Metro-Vick hauls 400 happy tourists to Courtmacsherry, escorted by a Henry Ford built blue and white Zodiac.

DRIMOLEAGUE TO BALTIMORE

Sometimes on our family excursions we got off at Drimoleague Junction and boarded the train to Baltimore. The train crossed the street at Drimoleague and then over the red steel 'Mutton' bridge to Madore's dark green-and-cream-coloured station. Woman power ran this beautiful and rustic station. The woman in question was the popular Jennie O'Leary, who used to paint Milepost 51 once a year. The railway ran next to the beautiful river Ilen for the next three-and-a-half miles before reaching serene Skibbereen at Milepost 57.

The Brunswick green and cream station was very extensive with a lot of closed and rusting buildings from the defunct Schull narrow-gauge railway line. Field's bakery provided us with creamy buns and sweet cake, which filled our food tank. Our train guard Timmy Donovan flagged our C class silver engine train to move out. Train driver Tags Tagney eased our engine C222 past the West Cork Hotel across the main street and through a very deep rock cutting named the Khyber Pass.

We then passed Milepost 53 and proceeded to Creagh Station in its mint green and off white colour scheme. This was a one-platform station, bereft of any siding. Miss O'Donovan was the halt keeper here and she had the trains, her radio and her small white pet dog to keep her company. A big black American Plymouth Fury car was parked down the untarred station lane. Grass grew in the middle of this laneway, often complete with daisies and daffodils. A very idyllic scene surrounded us as we observed the Church of Ireland church from our green carriage window with its four finials pointing upwards from its lovely stone tower – beautiful architecture, beautiful ambience.

Now we passed through a very rocky outcrop with plenty of bends and curves. A very old disused church and graveyard were passed before we reached Baltimore railway terminus. These buildings looked resplendent in their bright red colour scheme. Colour consultants were brought in from Denmark to advise the Irish railways on which colours to use.

All in all, looking back on these childhood journeys, I was privileged to see such outstanding scenery and interesting characters – all thanks to this great rail system.

RIGHT: Drimoleague looking west in splendid sunshine.

BEAUTY

Macroom Engine No. 5 was nicknamed 'Andrew'.

A symphony of shadows, steel and steam crossing Brian Boru Bridge on its journey from Dublin to beautiful Baltimore.

'The Green Goddess' – No. 1 route served Patrick Street, Albert Quay station and Blackrock.

The 'Ford' train brought Ford cars, vans, tractors, lorries and parts to stations nationwide. One Ford driver was known as 'Bunter the Shunter' (woe betide anyone who did not chain the vehicles properly).

A fine study of Albert Quay Bus and Rail Station. The perfect Cork city transport hub, which catered for bus, train, car, taxi, horse and dray, bike and last but not least, shank's mare (walking).

Horsepower at Skeaf.
A drawing by Chris Larkin.

EXCURSION
TO
BANTRY

SPECIAL TRAIN WILL LEAVE

AT: RETURNING AT:

Return Fare   **Third Class**

TRAVEL IN COMFORT BY TRAIN

Children under fourteen years, approximately half fare. Tickets are issued subject to the Bye Laws, Regulations and Conditions contained in the publications and notices of or applicable to the Board.

Fine sunshine lights the 12.15 p.m. Cork to Bantry train. Maurice 'Moll' Harrington is running to retrieve his luggage.

Cross City Rail 1961.

This broken train window was the result of the Billy Fury V Connie Lucey saga – because of Bishop Lucey's anti-rock and roll speech. The Togher Teddy Boys pelted the train with stones, because they thought that the bishop boarded this train.

View taken from the steps of the Albert Quay signal box. The Goulding's train is parked on the right.

The interior of railcar 2644 bound for Bantry, complete with its plush and comfy red and brown seats and Formica tables.

A rail enthusiast Jim O'Dea and friend under the Brunel designed roof at Albert Quay.

BELOW: Tickets issued down through the years – including the very last ticket issued in Albert Quay signed by the stationmaster, C. Twomey, on 31 March 1961.

A view taken of the departing train from the Hibernian Bridge known locally as the 'Cuckoo's Nest'.

No. 100 loitering without intent under the Hibernian road bridge. Her sister engine is now preserved in Downpatrick.

BELOW: I took this photograph inside the Rocksavage rail tunnel; it was famous for courting couples and bats. This is now the South Link Road, complete with polluting diesel-belching emissions from passing cars and commercial trucks.

Rocksavage tunnel.

The 'Clyde' cutting between Glanmire Road and Albert Quay station.

CIÉ ticket office with glass roof.
Well-designed images available from the booking office.

Hughie Harrington unloads a box of Rhode Island Red hens at Albert Quay.

Train passing St Finbarr's hospital, on the right note new sleepers, rails and ballast.

The Bantry bound rail car heading for the Snotty Bridge, gaining traction and speed.

Train at Togher.

Aboard the C222 crossing the Viaduct at 20 miles per hour speed limit.

The gateway to West Cork: the Chetwynd Viaduct on the Cork–Bandon Road.

Donie Murphy's Anglia slips under the viaduct past the Caltex petrol pumps

A German view of a Kinsale troop supply train. Between 1905 and 1960 countless German and Austrians filmed this fine line. This was the Mount Everest for road bowlers. Mick Barry lofted it in 1955, as did Hans Bohllen from Germany in 1985.

A view from the centre corridor coach number 1494 on Chetwynd Viaduct showing Dr Cronin's dispensary on the left. Also to the left, an ice blue Ford Zodiac on the wrong side of the road.

Misty and mystical Twomey's Castlewhite level crossing.

The train from Bantry – a painting by Chris Larkin.

A drawing by Chris Larkin of Waterfall booking office.

A view from Waterfall footbridge, Maurice Tobin has the fire lighting in the signal cabin.

*Inside the signal box with Maurice.
Note the beautiful pastel colours on the tongued and grooved timber panelling.*

Looking from the window of coach 1084. The ruins of Ballymacadane Abbey, which was founded in 1415 for the Augustinian nuns. Adjoining these ruins were the remains of an ancient Gaelic castle. This place was nicknamed Verdun after the First World War trenches. Oh yes that other puritanical gentleman Oliver Cromwell gave a courtesy call here during his campaign of shame.

Bus tours also started at Albert Quay bus and rail station.

Dinny Keane inspects a lady's ticket inside Goggin's Hill tunnel.

Approaching Goggin's Hill tunnel. The Pilgrim Footbridge, which was used by people going to mass at the mountain church nearby, flanks the Scottish lattice signal in the foreground. This red-latticed footbridge was sometimes called the Pilgrims Perch. The Togher and Ballyphehane Teddy boys often walked the line to here and also Togher schoolboys like John O'Connell, Davy Moran, Noel O'Meara, Brendan Murphy, and Jimmy Bantry Keohane often went there. Our verbal greeting was 'Howdy Pilgrim' in a John Wayne accent.

East Cork stone adorns this splendid portal in Ballinhassig.

The Crossbarry signal box and footbridge. This was a former junction serving Kinsale.

Joe St Leger stands on the right hand platform at Ballinhassig station looking resplendent with its new coat of paint.

Crossbarry as viewed from carriage 1450, Jim Lane (station master) and Dinny Keane survey the scene.

Cormac O'Donovan on the up platform. Unfortunately Upton was the scene of a terrible massacre. On 15 February 1921 having received good information that crown troops would be travelling on the 9.15 a.m. down train from Cork, the local flying column decided to attack this train which included many innocent passengers. Volley after vicious volley of lead tore into the timber carriages. What the flying column did not know was that one of their officers tipped the British off. Did he do it because of the women and children on board? Or did he do it because of the money? We will never know. Nine coffins were the result of this sad debacle. John Bennett who later worked with my grandfather, Jerry Cullinane, at Timoleague station said this scene was pure evil. A local informer swung the balance for the crown on this occasion.

Upton station.

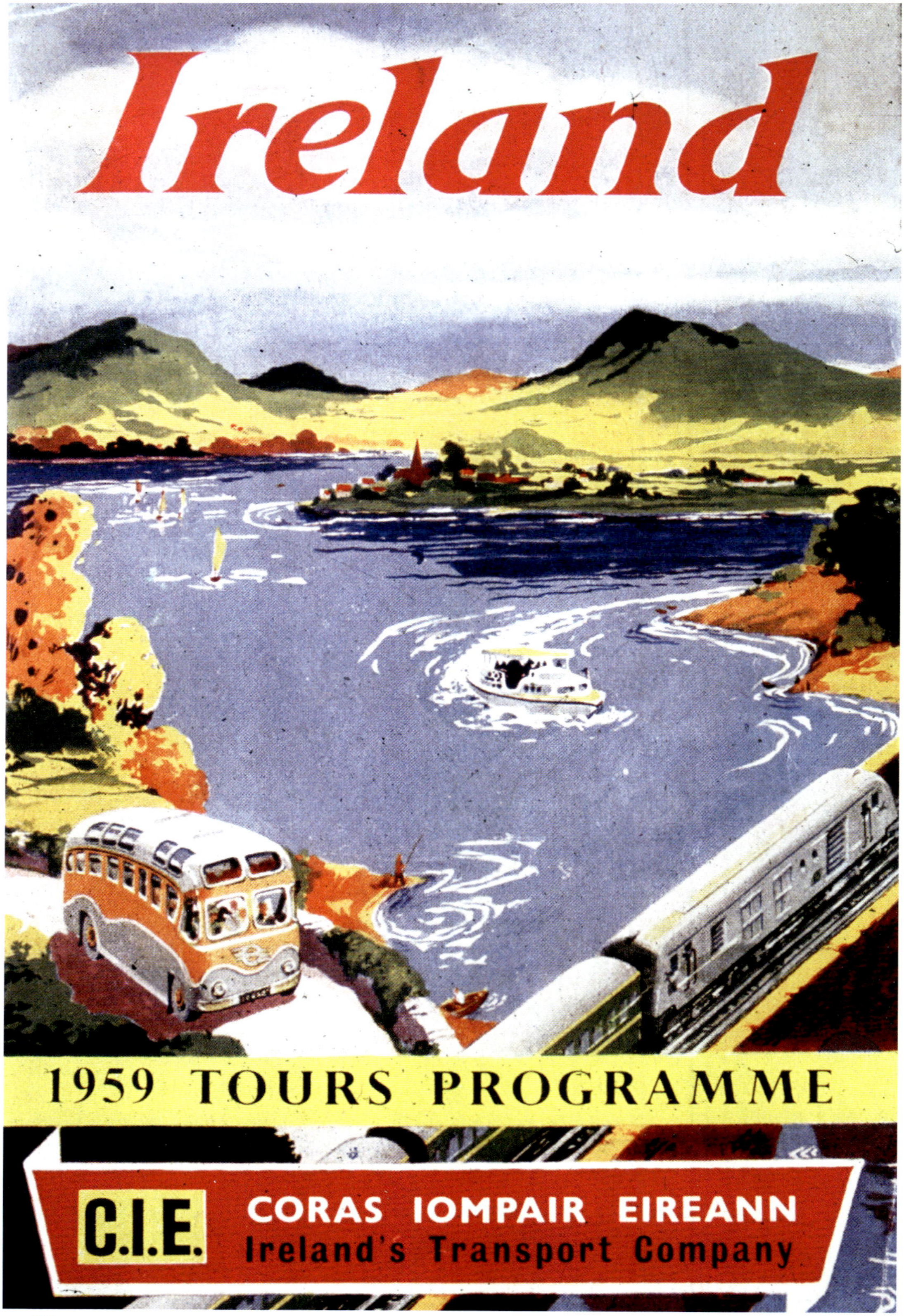
Ireland
1959 TOURS PROGRAMME
C.I.E.
CORAS IOMPAIR EIREANN
Ireland's Transport Company

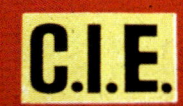

SUNDAY, 5th JULY, 1959
EXCURSION TO BANTRY

Cork Junior Football Championship:
DOHENYS v. SKIBBEREEN

		P.M.	RETURN FARES 2nd CLASS TO BANTRY	
			£	D.
Cork	dep.	12.25	11	0
Waterfall	,,	12.40	9	9
Ballinhassig	,,	12.49	9	3
Crossbarry	,,	12.56	8	6
Upton	,,	1.01	8	0
Bandon	,,	1.09	7	0
Clonakilty Junct.	,,	1.18	6	3
Desert	,,	1.26	5	9
Ballineen	,,	1.31	5	0
Dunmanway	,,	1.43	3	9
Bantry	arr.	2.25	—	

Returning from Bantry at 7.00 p.m.

IN ADDITION, CHEAP TICKETS WILL BE ISSUED FOR TRAVEL TO, FROM AND BETWEEN ALL STATIONS SERVED BY THE ABOVE TRAIN
CHILDREN UNDER FIFTEEN YEARS APPROXIMATELY HALF FARE. Tickets are issued subject to the Bye Laws, Regulations and Conditions contained in the Publications and Notices of or applicable to the Board.

50/6449

TRAVEL in COMFORT and save time

C.I.E.

CORAS IOMPAIR EIREANN

*A black and white image of the Innishannon tunnel.
Note the tall telegraph pole – there was life before the mobile phone.*

The Innishannon tunnel portal in the foreground with brand new rails, etc.

A view from carriage 1902 crossing the Innishannon red latticed viaduct.

One man and his bicycle in Bandon station.

Underneath the Cork Gas Works wall a Bandon loco awaits its call.

A Munster Hurling Final GAA sports train at Bandon awaits the whistle.

A view of Clonakilty junction from carriage 1906. We see Herr Lammers in his plus fours. Donie O'Donovan is looking after the parcels.

Delightful Desert station with its lovely privet hedge was a favourite place for birds' nests and bumble bees.

Ballineen station – the Beara Mountains far away in the distance await us.

Dunmanway with Atkins' Mill standing tall.

An RIC officer once asked Michael Collins on this very footbridge: 'what is your function on this junction?
Mick replied, 'your malfunction'.

Delightful Drimoleague.

The red flag on this train signals that another is to follow.

Drimoleague looking east. Barry Deane has re-erected this GSR sign nearby for the pleasure of future generations of locals and tourists alike.

Drimoleague junction, with Baltimore branch to the left and Bantry mainline to the right.

Aughaville.

One girl and her bike alight at rustic Aughaville with its Cedar of Lebanon trees lining the gravel platform.

Denis Keohane steps down from Bantry signal box.

This is not a log cabin in Montana but delightful Durrus Road.

A view from railcar 2641 at Durrus Road.

A 1961 sunny scene at Durrus Road.

Bantry signal at the down position.

The Glengarriff bus at Bantry awaits the arrival of the train. The 1957 Cheverlot in front of the bus belonged to the 'Canadian'.

A rain polished platform.

End of the line in Bantry, under the spell of the Beara mountains.

Eccles's exotic hotel with passengers sailing towards her in the foreground.

Puxley's Palace, Castletownbere, once the seat of O'Sullivan Bere.

A fine view of Glengarriff pier with a horse drawn hackney passing the Baltic pine fencing. The Cork Bandon and South Coast railway building survives to this day, well worth a visit.

The Pier Glengarriff Co. Cork.

Amazing Allihies.

An old copper mine steam engine rusting in peace.

A beautiful sight, one of the mining engine houses – mining material came to Bantry by rail from all parts of the country and transhipped by road for the Allihies mines.

The road from Allihies to Urhan — another mining area.

The road to Eyeries which is still untarred.

Ballinascarthy junction with the Courtmacsherry line to the left and the Clonakilty line is to the right. The local creamery now operates these preserved buildings.

The Irish Railway Record Society train at Ballinascarthy 1960 with Roy Hammond on board. Claude Fogerty is on the platform looking smart in his Prussian officer style classic clothing.

A newly painted station in Ballinascarthy.

Ballinascarthy junction, with Cork line to the left and Courtmacsherry to the right.

An open flat wagon No. 10 M at Bandon delivering a black mark II Ford Consul to a Mr Hodnett. The Ford logo was and will always be a design classic. Ford gave short shrift to vested interests like advertising agencies urging him to upgrade his logo.

Passing Mounteen Castle aboard green loco 231, near the old centre of Tuath Mountain up to the Glen Road. It was a MacCarthy castle and the local IRA flying column stashed Luger and Mauser pistols, ammunition and Lee Enfield rifles here during the War of Independence.

Skeaf name board is now preserved in Dublin. Chris Larkin with parent on platform, eagerly awaiting the pleasant and picturesque trip to Courtmacsherry village, woods, strand and a daytime public house ceili session.

Chris Larkin and parents at Skeaf alight from carriage No 1088 for a day's outing to the pretty village of Clogagh, now but a shadow of its former self, as are the majority of Ireland's rural villages.

BELOW: My grandfather (top left) with other railway workers at Timoleague station.

A tranquil Timoleague scene.

An India ink drawing by Chris Larkin – 'The Abbey'.

Timoleague summer 1960 the ozone layer is unbroken and happy.

Courtmacsherry.

An India ink drawing by Chris Larkin.

Wagons on Courtmacsherry pier.

Clonakilty.

A J.F. muck spreader awaits collection from the freight platform at Clonakilty.

Shadows to the fore at this Clonakilty Japanese style building.

Madore station is very reminiscent of Scottish highland scenery and stations. If you ever want a flavour of what the West Cork lines were like, go to Queen's station in Glasgow, board the train to Mallaig and see for yourself – it's a beauty.

Near Madore.

An engine at Skibbereen.

One man and his dog.

Jack Riordan at Skibbereen, March 1961.

Baltimore station with posters.

Baltimore station.

Con Dan Kellegher and Tom Dineen observe a well tailored tourist at Macroom terminus.

Bantry terminus.

THE DESTRUCTION OF THE RAILWAY

The West Cork lines had survived the Great Famine, the First World War, the War of Independence, the Civil War, the 1929 depression, the 1930s recession, the Second World War, the 1947 coal famine and the mid-1950s slump.

In the USA and Europe during the 1950s the vast and financially powerful road lobby finally made their move against the railways. This was disastrous, particularly since we now know that railways, especially electric-powered ones, are planet earth's best friend, protecting the ozone layer and helping to keep the greenhouse gas effect at a low level.

The German chancellor, Herr Adenauer, gave the road lobby short shrift on their Teutonic tonnage thrust; in Germany it was *bahnhof uber alles* (Railways above All). However, the Irish road lobby had been putting the political parties under pressure since the 1930s. W. T. Cosgrave kept them at bay for a long time, but this powerful lobby worked very hard to curry favour with politicians for their own agenda. Jack Lynch told me after the opening of the Henry Ford GAA pitch in Ballinascarthy back in the 1980s that there were a few cabals in that late 1950s Fianna Fáil administration. Jack also told me some chilling stories of their endgame, which was the total change-over of people and goods from rail to road.

During July 1957 a copy of the submission of the board of CIÉ was handed to the Beddy enquiry for its perusal. In this submission, T. C. Courtney's board asked all the relevant questions, while at the same time providing sound, sensible and solid advice on future Irish railway practice, such as dealing with illegal road haulage, state subsidies and the setting up of a central highways authority.

Being an astute and able manager, T. C. Courtney heard of impending moves being made against the railway system. As former haulier Timber Thady Twomey told Colm Creedon, the dogs on the road knew of the plan to switch the transportation focus from railways to roads for financial gain.

Only two members of Beddy's five-person committee were sympathetic to the railway system. As a result, Beddy rejected the CIÉ submission. However, Seán Lemass then rejected the Beddy report – but there was a sting in the tail, as he gave CIÉ unlimited powers to close some mainline, secondary and branch lines when he introduced the Transport Act of 1958. This proved to be a death sentence for a lot of railways.

The 'Save Our Railways Association' was formed in Cork city on 8 May 1957 due to discreet tip offs by Fianna Fáil TDs, such as Seán MacCarthy and Martin Corry, etc. As the road haulage lobby had now become a powerful pressure group, the Save

Our Railways Association's first job was to compile a solid and sensible fact-driven memorandum, putting forth honest arguments in favour of preserving all the railway lines running in and out of Cork.

A SURVEY OF PUBLIC TRANSPORT REQUIREMENTS IN WEST CORK.

-ooOoo-

Compiled by the

SAVE OUR RAILWAYS ASSOCIATION.

-ooOoo-

INTRODUCTION.

Following upon the completion of our general memorandum entitled "Proposals for Development of C.I.E. Rail Services", we, the Committee of the Save Our Railways Association, now embark on the production of a more specialised document involving proposals for the economical development of a specific section of the C.I.E. Rail Network.

We should emphasise, at the outset, that our choice of section has not been prompted by any feelings of local sentiment towards the West Cork Area. In view, however, of its proximity to the headquarters of our Association, many of our members are conversant with the transport requirements of the locality chosen and are therefore in a very good position to comment on them. Further, persistent local rumours of impending closure have been current in recent months so that the time is now opportune to review public transport in West Cork before the present unsatisfactory position deteriorates.

Despite formidable competition from road transport, and the decline in the population of the area, the Cork and Bandon railway line still forms the backbone of transport services in West Cork, and retains its importance as a vital "lifeline", linking the local towns and production areas with Cork City and, through it, the rest of the Country. Far from being a mere 'rural byway', the West Cork line is an important Secondary Rail Route, connecting the second City of the Republic with Skibbereen (Capital of West Cork), Bantry (Gateway to Glengarriff), and serving the important towns of Bandon, Dunmanway and Clonakilty, together with Timoleague and Courtmacsherry, centres of the richest beet growing area in the South of Ireland.

We welcome the re-assuring references to subsidiary railway lines in the policy statement recently delivered by the Chairman of C.I.E., particularly the declaration that the Board are not going to close any line where there is any hope that their best efforts can save it. We are firmly of the opinion that the West Cork line comes within this category, and we are convinced that it will respond successfully to a scheme of development and re-organisation, such as we have drafted in the following pages.

We hope that the Board of C.I.E. will give active consideration to the contents of this Memorandum, particularly the proposals and recommendations contained in Part 2., as the future prosperity of the entire West Cork area must depend, to a large extent, upon the availability of an efficiently operated local network of rail and feeder road services.

J. St. Leger.)
)
C. Creedon.)
) Committee.
W. McGrath.)
)
J.E.A. Cusack.)

Cork. 23rd May, 1959.

Intro to A Survey of Public Transport Requirements.

The Bantry line had an advantage over the other Cork lines, as it was both a suburban and rural line. It was also a tourist and heritage line, serving Ireland's most scenic region for over a hundred years. Walter MacGrath told me years later that around this time (August 1960) it finally dawned on the Save Our Railways Association that if T.C. Courtney had been able to close the branches off the Bantry mainline, as he'd proposed back in the early 1950s, the Bantry line would have been safe from closure. To quote Joe St Leger: Courtney should have been allowed to prune the branches, thus saving the trunk route.

The new Parnell Place bus station in Cork was built by P. J. Hegarty and opened on 12 October 1960 at a total cost of over £70,000. Albert Quay station had been the main bus station during the construction period of 1958–60. Walter MacGrath always maintained that the two-year costs of running the buses at Albert Quay, charged to the West Cork Railway, should have been deducted from the railway account of the period. This would have helped to swing the balance in favour of the retention of the West Cork lines.

Despite the mounting financial issues, about this time all the stations were repainted, fencing replaced, roofs replaced, new track, sleepers and ballast laid. An enormous amount of taxpayers' money was spent on new name boards, lighting, and the upgrading of platforms, bridges, tunnels, etc. Surplus uniforms, caps, overalls, boots, and tools were issued to all and sundry. Barrels of high-quality paint appeared out of nowhere, and there was, I believe, unwarranted duplication of posters, and the introduction of chrome winged-wheel cap badges that unnecessarily replaced the older copper issue. On top of all this unnecessary expenditure, people also began to find it difficult to charter trains on the West Cork system; for example, the cancellation of the proposed ballroom excursions for the Lilac dancehall in Enniskeane and the GAA found it hard to charter trains. Similarly, hundreds of passengers were left stranded on platform stations on countless occasions.

It didn't help that T. C. Courtney was ousted from his position around this time, replaced with Todd Andrews, with the backing of a small cabal in Fianna Fáil. Thus began CIÉ's darkest period of all from 1959 to 1966 when dissent, strikes and low morale were the order of the day.

During this period, the winged-wheel logo was replaced by the unpopular broken-wheel logo at an enormous cost to the taxpayer. Even worse, the beautiful bright green livery used on CIÉ trains was replaced again at a very high cost. The dreary dark and drab livery – some would say anti-livery – of black and tan colours made things far worse. The classic two-tone green bus livery was replaced by a very dark blue and

1960, people left behind in Albert Quay.

SUNDAYS, 2nd, 16th & 30th AUGUST 1959

EXCURSION TO COURTMACSHERRY

			Return Fares, 2nd Class To Courtmacsherry	
			S.	D.
Cork (Albert Quay)	dep.	11.30 a.m.	7	3
Waterfall	,,	11.45 ,,	6	0
Ballinhassig	,,	11.54 ,,	5	6
Crossbarry	,,	12.01 p.m.	4	6
Upton	,,	12.06 ,,	4	3
Bandon	,,	12.14 ,,	3	6
Clonakilty Junct.	,,	12.24 ,,	3	0
Ballinascarthy	,,	12.41 ,,	1	9
Timoleague	,,	1.16 ,,		6
Courtmacsherry	arr.	1.35 ,,		

Returning from Courtmacsherry at 8.15 p.m.

Children under 15 years approximately Half-fare.

Tickets are issued subject to the Bye Laws, Regulations and Conditions contained in the Publications and Notices of or applicable to the Board.

Coras Iompair Eireann

50/9535

cream, which became very dirty quickly. The beautiful bus signs with the Gaelic font were also replaced with those of lesser quality design. The very high standard CIÉ posters, issued between 1945 and the late 1950s, were also replaced by lower-level designs.

As previously mentioned, after 1959, a lot of money was wasted on needless cosmetic items; one could call it a spending spree. One example being that every single bus stop metal sign was replaced by a round metal one. This expenditure alone could have paid for the moth-balling of the Cork-to-Bantry mainline until its resurrection – like the Midleton line, which is doing so well today. A resurrected Bandon line would easily double the Midleton line figures – it's not all about figures, of course, but we are much more aware of the need to sustain the survival of planet Earth for the next generations.

During the 1960–66 spending spree, £151,000 was spent on the acquisition of road vehicles to replace closed branch lines, which would have been more than enough to revive the railways. From 1960 on £6,225 was expended on the mezzanine floor of Busáras Dublin. As already mentioned, the construction cost of the Parnell bus station Cork was £70,000. On 29 June 1961, £9,885 was spent on staff cars and vans. The passenger platforms at Albert Quay were converted at a cost of £5,500 in 1963. From 1964 on £425,790 was the budget for new road vehicles. Unnecessary livery changes did little to enhance the corporate image. The railway station platforms with their traditional cut stone slabs were ripped up and replaced with expensive black and white tiling. The annual subvention money given by the government to operate the railways was misspent on all of the above items.

A policy of reducing flat wagons on the rails prevented the timely delivery of Ford Zodiacs, Zephyrs, Consuls, Anglia's and Thames trader vehicles to all parts of Ireland. Less wagons were also available for the carriage of beet to Mallow factory, thereby benefiting the private haulage sector. The proposed early morning Bandon-to-Cork train to serve Ford's, Dunlop's and Goulding's was put on ice, as was the proposed use of more powerful engines on heavy-freight trains to and from Bantry. The continuous breaking down of the Metro-Vick loco units added to running costs and unreliability of the railway. Colm Creedon, suspicious by nature, asked Walter McGrath, 'Is this a plot or is it pure bad luck?' John Waters wrote an interesting article on Irish railways on 31 August 2014 in the *Irish Independent*. A former CIÉ man, he told us of the anti-rail economists, journalists and transport experts who were given lunches and junkets by the road lobbyists.

Ray Good was working as usual in Bandon on Tuesday 27 September 1960. John King and others were busy painting the station windows, doors and walls when Cork district Superintendent P. J. Herbert came in and assembled everyone. There was no tension in the air as all and sundry expected the announcement of expanding morning and Sunday services. But worry began with the shock announcement that they were to close the entire West Cork rail system as and from 31 March 1961. Just like that, the railway was no more.

A GREEN FUTURE

Life has gone on without the West Cork railways. It had to. The line was destroyed very rapidly and so avoided EEC entry in 1973, which would have preserved the railway right of way for future generations. Motorways now rule, even though each new road built increases the amount of traffic, resulting in severe congestion during the morning and evening peak hours. People now loudly shout 'bring back the train' when stuck in traffic, and it seems one answer is electric railways and trams.

Green is the buzz word used by everyone today and rightly so. And so hopefully trams and light rail will make a comeback in the greater Cork area, much like the closed Harcourt Street line did in Dublin, reopening as the Luas. Electric trams and light rail certainly seem like the way forward. A light rail system westwards to Ballincollig and Ovens will make a good start. And, who knows, maybe West Cork will be next.

The population of Ireland deserve better: *Aber Bandon wird sich erholen* (Bandon will Rise Again)

Albert Quay carriage at Youghal terminus.

APPENDIX 1

POEMS & ART

by

Chris Larkin

ODE TO JOE ST LEGER

Joe St Leger was a railway photographer of great renown,
Known in Belfast Station, Kingsbridge and Bantry town
With his brother John and Colm Creedon too
Every station visited by railway men through and through.

On the train to Bandon they oft be seen
Seated in the railcar, in its bright shade of green
He got where he went by road and rail
Chasing locos and coaches complete with the flying snail.

His slide shows covered Britain to beautiful Bantry Bay
With a merry sense of humour he always carried the day
Whether to Ray Good's home in Gaggin or Alan Barry's Ramble Inn
Joe would often ring me saying lets go there once again

Joe's mild manners were a joy to behold
His knowledge of railways a story untold
He didn't like busmen
Except Joe Lawton and Ritchie Lee

He called them the opposition
Preferring trams near the river lee
Now Joe has left us as legends often do
Never to be forgotten by friends old and new.

POETRY IN MOTION – NUMBER 464

Within the rugged Rocksavage Tunnel,
A Bandon tank emits smoke from its funnel,
The smoke is a plume of grey and white,
A seamy sight in the still of night.

And now with incurious stars,
Gleam on green hurtling cars,
Against the blind and awful rain,
In darkness this our West Cork train.

In dark defiance barks and shouts,
This noisy Cork rebel pouts,
464 passes on and leaves no trace,
For darkness holds its hiding place.

Serene and beautiful this loco king,
Steam and steel make it a living thing,
Our houses lie happily pure and still,
In Doughcloyne, Chetwynd and Spur Hill.

Our train intensifies the dark,
As we pass Charley Hurley park,
Quietness holds our weary feet,
We clank and clang through the sleet.

Through Desert railway station,
No green door will open wide,
Not an eye will ever see,
Who this train traveller may be.

Upon my crimson coloured seat
That radiates some happy heat,
Feeling natural and mighty keen,
Rolling in to Bantry Town, cool and clean.

The station lamps shine white and sound,
I descend from the train to hallowed ground,
The train ticket was not that dear,
Bless the loco that got me here.
A Chrysler taxi takes us to Castletownbere,
Where we witness merry fishermen on the tare,
This soft pillow now takes over,
Time to dream of my Collie dog Rover.

Goggins Hill tunnel – a drawing by Chris Larkin.

IN THE TRAIN

While sitting in the West Cork train,
I like to look around,
And with some discretion,
Study those that are city bound.

The regulars will always,
Take the same old seat,
When it's occupied,
They bade a quick retreat.

Behind me is a loud lady,
How old I cannot say,
Whose views and verdicts,
Are broadcast all day.

While to the front I see,
Our champion fumbler there,
While she is searching for her cash
A victim pays her fare.

From waterfall we weave our way,
Up on Chetwynd Viaduct most high,
Is it not the perfect day
Look at that bright blue sky.

Passing under Meagher's Lane Bridge,
Onward to Milepost 4,
Rolling past Doughcloyne,
We witness a bowling score.

Our train guard is Dinny Keane,
A merry old soul is he,
He cheerfully announces,
We will soon be at Albert Quay,

He opens the train door,
Allowing a Lady to alight,
She smiles and blows a kiss,
Much to Dinny's delight.

With this joyful journey over,
There is nothing left to say,
But eat a tube of Smarties,
And be on my way.

The China Wall – a painting by Chris Larkin.

THE BRIGHT BEARA WAY

Bright Beara Way,
Skies of heavenly hue,

Shine over Allihies and Urhan,
The whole day through.

Night will never diminish,
Your sweet celestial blue,

Land of magical mountains,
With turbulent tides on cue.

Realm of enchantment,
Won't erase your mystical spell,

Ever and always,
Your oracle will dwell.

When our afterlife will come,
And come it certainly will,

Will the valleys of Beara,
Give us the same cosmic thrill.

From the fish in the sea,
To the Collie dog on the road,

We will have to unburden,
Our complete earthly load.

Upton – a painting by Chris Larkin.

APPENDIX 2

RAILWAY STAFF

Years ago, when Colm Creedon was urging me to do a West Cork railways book, he gave me a list of people who worked on this wonderful railway from Cork to Beara:

Albert Quay Railway Station	
C. Twomey	Stationmaster
J. Smith	Traffic Inspector
P. Hennessy 'The Beagle'	Guard and Station Porter
Chris Beakey	Signalman
Mike Killackey	Inspector
Batt 'Rising Sun' Foley	Ganger
Danny Cullinane	Goods
Dinny Hannigan	Train Guard
Jackie 'Sprat' Ahern	Train Guard
Bill Holland	Train Guard
Tommy Philpot	Train Guard
Tim McCarthy	Train Guard
Dick Forde	Train Guard
Tom Gosnell	Train Guard
John 'Spud' Murphy	Train Guard
Dermot Linehan	Train Guard
Dinny Buckley	Train Guard
Jimmy Collins	Train Driver
John Collins	Train Driver
Billy Cogan	Train Driver
Patrick Meighan	Train Driver
Jacky Ryan	Train Driver

Thossy O'Donovan	Train Driver
Jerh Canty	Train Driver
Mick Minehane	Train Driver
Jer Long	Train Driver
Jackie Leahy	Train Driver
Jack Buckley	Train Driver
Michael Quirke	Train Driver
Batt Sheehan	Train Driver
Bill Cogan	Train Driver
Dick Fitzgerald	Train Driver
Finbarr Daly	Train Driver
Tom O'Connell	Train Driver
John Cahill	Train Driver
Leo Delaney	Train Driver
Tom Ryan	Train Driver
Batt O'Brien	Fireman
Denis O'Sullivan	Fireman
Bertie Walsh	Fireman

Waterfall Station	
Maurice Tobin	Halt keeper and Signals
John Joe Crowley	Line Ganger

Ballinhassig Station	
Dan O'Mahony	Halt keeper and Signals

Crossbarry Station	
Jim Lane	Halt keeper
Tim Murphy	Signalman

Upton Station	
Cormac O'Donovan	Halt Keeper
William Barry, Snr	Ganger
William Barry	Ganger

Bandon Station	
Mr O'Brien	Stationmaster
Dick O'Regan	Train Driver
Connie Walsh	Signalman
Tommy Desmond	Signalman
John Sisk	Porter
Ray Good	Clerk
Ted Desmond	Ganger

Clonakilty Junction	
Donie O'Donovan	Halt Keeper
Michael White	Ganger
John Flynn	Ticket Man
Michael O'Donovan	Signalman

Desert Station	
Mrs Duckett	Halt Keeper

Ballineen Station	
Mr Curtain	Stationmaster
John O'Brien	Signalman

Dunmanway Station	
Mr Corkery	Stationmaster
C. Hennessy	Signalman

Drimoleague	
Mr Bradley	Stationmaster
Mick Minehane	Signalman
Tim Crowley	Signalman
Michael O'Connor	Head Porter

Aughaville	
Mrs Horgan	Halt keeper

Bantry Station	
John O'Donovan	Stationmaster
E. Tagney	Train Driver
Jack 'Manifold' Daly	Train Driver
D. 'Boots' Murphy	Train Driver
Arthur 'Attie O'Connor	Signalman
John Murphy	Porter
Dinny Keane	Train Guard
Donie Collins	Train Guard

Ballinascarthy Station	
Claude Fogarty	Stationmaster
J. Mullane	Signalman
J. Foley	Ganger
Con O'Donovan	Ganger

Timoleague Station	
John O'Donovan	Halt Keeper
Jerry Cullinane	Goods

Courtmacsherry Station	
Paddy Madden	Halt Keeper

Clonakilty Station	
William Patterson	Stationmaster
Timmy Donovan	Signalman
D.J. McCarthy	Porter
Bill Lombard	Train Driver
Dick Nagle	Train Driver
Christy Googan	Train Shunter
Dan 'Tailor' O'Donovan	Train Guard

Madore	
Janie O'Leary	Halt Keeper
Con O'Leary	Ganger

Skibbereen Station	
Mr O'Rourke	Stationmaster
C. Tagney	Train Driver
Mick McCarthy	Ganger
Denis O'Donovan	Signalman
Mr Hanley	Porter
Gerry Calnan	Goods

Timmy Donovan	Train Guard
Dan Dineen	Train Guard
Creagh Halt	
Miss O'Donovan	Halt Keeper
Baltimore Station	
Michael O'Donovan	Halt Keeper
Charles Davis	Ganger

During the Cork Bandon & South Coast Railway, Great Southern Railway and the Córas Iompair Éireann era, many railway workers positions were temporary, hence a lot of people like porters, plate layers, gate keepers, goods men, etc., are not on any lists or books. I contacted the trade unions but they were not very helpful to my transport and social history research and unfortunately many people associated with the railways cannot now be named.

CÓRAS IOMPAIR ÉIREANN

Account — Period Ending FEB. - 1960

To EMERALD ISLE MINING Co ALLIHIES Co CORK from IRISH OXYGEN LTD WATERFALL

Please Remit to COLLECT AT BANTRY RAILWAY DEPOT WITHIN FOUR DAYS

PLEASE QUOTE CREDIT

Cheques to be made payable to CÓRAS IOMPAIR ÉIREANN and crossed " & Co. ".

Details of Charges are shown on attached Goods Progress Notes.

PLEASE REMIT THE AMOUNT DUE £ £20 : 6５ : 0

Date	Reference No.	£	s.	d.
11-2-60	GAS REFILLS	£20	6	0
	OXYGEN CYLS—			

APPENDIX 3

SIGNALLING, CARRIAGES

There were rules for working the single line between Cork, Bandon, Kinsale, Baltimore, Bantry, Courtmacsherry and Clonakilty:

The Colours of Staffs
Metal staffs were necessary safety devices used to ensure that trains did not crash on single rail tracks (and are not necessary on double tracked lines) and these have been replaced by electric railway traffic lights. Between every station the colours were different and trains could only proceed if the signal sign is horizontal.

Cork to Macroom Junction	White
Macroom Junction to Waterfall	Green
Waterfall to Ballinhassig	Dark Green
Ballinhassig to Kinsale Junction	Mid Blue
Kinsale Junction to Bandon	White
Bandon to Clonakilty Junction	Red
Clonakilty Junction to Desert	Blue
Desert to Ballineen	Red
Ballineen to Dunmanway	Blue
Dunmanway to Drimoleague	Red
Drimoleague to Durrus Road	White
Durrus Road to Bantry	Red
Clonakilty Junction to Ballinascarthy	White
Ballinascarthy to Courtmacsherry	Yellow
Ballinascarthy to Clonakilty.	Red
Drimoleague to Skibbereen.	White
Skibbereen to Baltimore.	Red
Kinsale Junction to Kinsale	Yellow

A coloured metal staff had to be carried on each train to and fro and no train was allowed to start without one of these from the above stations. This was due to safety reasons.

Shape of the Staff	
White	Oval
Green	Hexagon
Red	Round
Blue	Square
Yellow	Three sided
Cork & Bandon Railway 1872 Regulation Book	

Macroom Junction Signal Instructions
Engine drivers approaching from Macroom are to give one distinct whistle.

Engine drivers approaching from Bandon are to give two distinct whistles (every other junction had a platform).

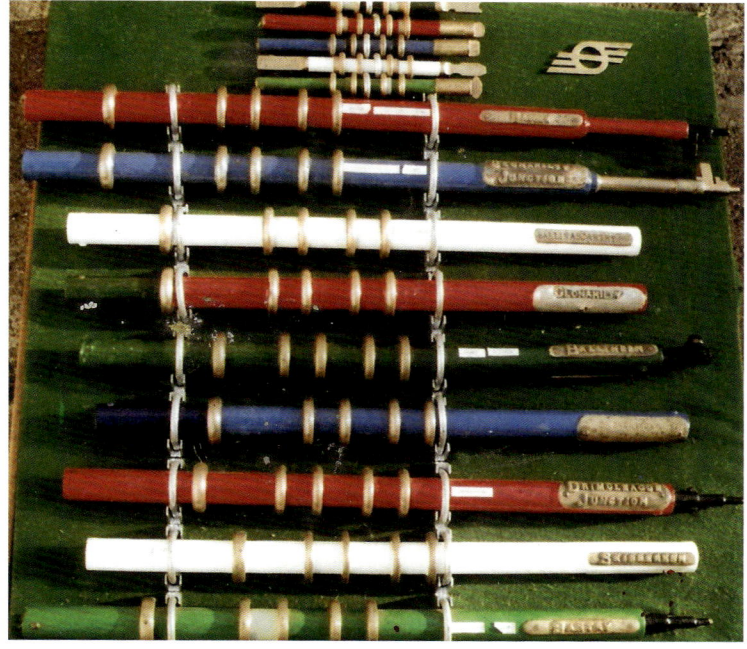

West Cork System Metal Railway staffs.

List of the timber and brick Cork Bandon & South Coast Signal Cabins 1899			
Place	Levers	Points	Signals
Albert Quay	40	20	20
Waterfall	12	5	7
Ballinhassig	10	4	7
Kinsale Junction	30	15	17
Bandon	20	8	12
Ballineen	12	6	6
Dunmanway	18	8	10
Drimoleague	31	16	15
Bantry	18	10	11
Ballinascarthy	17	5	8
Clonakilty Junction	27	12	13
Skibbereen	17	5	9
Madore	4	0	4
Knockbue	4	0	4
Farran Galway	10	4	6
Kinsale	10	4	6

Cork and Bandon Railway lamps from the nineteenth century.

Cork Bandon & South Coast Railway Carriages

No. Class	Date	Length	Wheels	Seats	Withdrawn
1 Third	1905	43 feet	8	80	1957
2 Third	1905	43.3" feet	8	80	1957
3 Third	1905	43 feet	8	80	1957
4 Third	1905	43.3" feet	8	80	1957
5 Third	1880	27.6" feet	4	40	1938
6 Third	1906	46.6" feet	8	50	1948
7 Third	1881	27.6" feet	4	40	1930
8 Third	1896	33.feet	8	60	1938
9 Third	1881	27.6" feet	4	50	1941
10 Third	1865	27.6" feet	6	50	1948
11 Third	1865	27.6" feet	6	50 rebuilt	1922
12 Third	1890	24.9" feet	6	50	1938
13 Third	1865	27.3" feet	6	50	1892
14 Third	1865	27.3" feet	6	50	1882
15 Third	1865	27.3" feet	6	50	1920
16 Third	1865	27.3" feet	6	48	1913
17 Third	1890	26.10" feet	4	40	1947
18 Third	1890	27.4" feet	6	50	1948
19 Third	1890	25.10" feet	6	40	1922
20 Third	1878	25.11" feet	6	20	1959
21 Third	1878	25.11" feet	6	20	1955
22 Third	1878	25.11" feet	6	20	1955
23 Third	1890	32.10" feet	6	60	1929
24 Third	1865	27.4" feet	6	50	1955
25 Third	1894	33.0" feet	8	60	1957
26 Third	1896	33.0" feet	8	60	1941
27 Third	1896	33.0" feet	8	60	1948

28 Third	1907	46.6" feet	8	50	1955
29 Third	1904	27.4" feet	6	50	1923
30 Third	1904	27.5" feet	6	20	1949
31 Third	1905	38.0" feet	8	40	1957
32 Third	1872	25.0" feet	4	40	1940
33 Third	1906	29.4" fee	6	50	1923
34 Third	1888	30.6" feet	6	50	1940
41 First	1879	26.6" feet	6	32	1949
42 First	1863	24.7" feet	6	32	1928
43 First	1891	34.0" feet	8	40	1957
44 Mixed	1906	46.6" feet	8	50	1953
45 First	1894	34.0" feet	8	40	1957
46 First	1899	33.10" feet	8	40	1949
47 First	1899	33.10" feet	8	40	1949
48 First	1876	27.4" feet	6	32	1943
49 First	1879	26.11" fee	6	32	1949
50 First	1902	37.10" feet	8	32	1923
51 First	1879	26.11" feet	6	32	1949
52 First	1892	33.11" feet	8	40	1957
53 Mixed	1908	48.11" feet	8	46	1959
54 Mixed	1901	37.10" feet	8	42	1959
55 Mixed	1895	34.10" feet	8	26	1960
56 Mixed	1908	48.11" feet	8	46	1959
57 Mixed	1901	37.10" feet	8	42	1959
58 Compo	1876	30.60" feet	6	36	1923
59 Mixed	1876	30.60" feet	6	36	1953

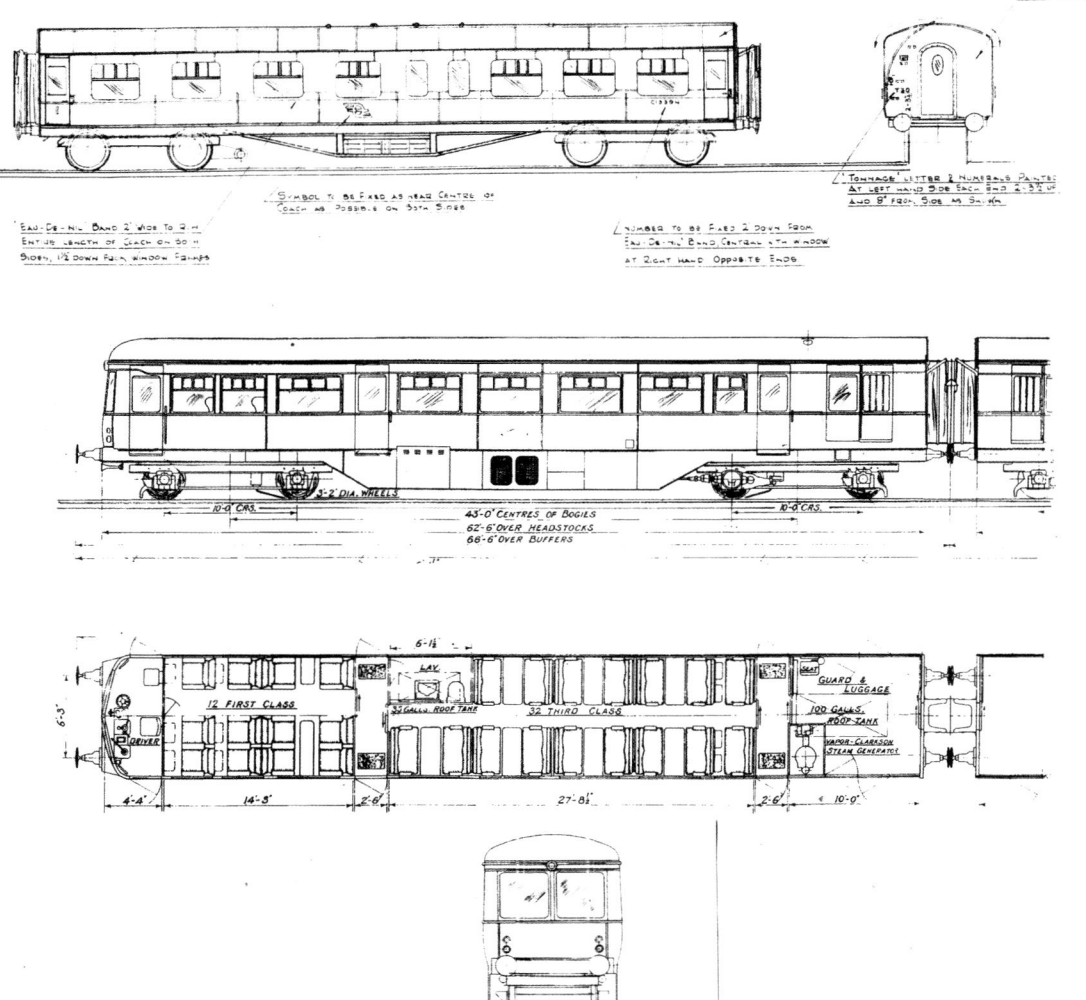

From 1925 onwards the GSR sent dozens upon dozens of different style carriages on to the Bandon system. MGWR (Midland Great Western Railway), Great Southern and Western Railway, The Great Northern, UTA (Ulster Transport Authority) all became part of to the system. Great Northern wagons, open wagons, brake vans, etc. – in fact every type of wagon available – was utilised.

APPENDIX 4

CORK AND BANDON STEAM LOCOMOTIVES 1849–1888

CORK BANDON & COAST RAILWAY LOCOMOTIVES (1888–1925)

The following is a list of the various steam locomotives provided by these two companies up until the formation of the Great Southern Railways in 1925. When engines were replaced or rebuilt the same numbers were retained.

No.	Name	Maker	Year	Withdrawn
1	*Rith Tineadh* 5	Adams	1849	1868
2	*Sighe Gaoithe* 6	Adams	1849	1868
3	*Fag an Beallach* 2	Tayleur	1849	1891
4		Tayleur	1851	1889
5		Sharp	1852	1887
6		Sharp	1852	1879
7		Fairburn	1862	1897
8		Bury	1865	1874
2		Bury	1866	1877
8		Bury	1871	1877
8		Bury	1871	1874
1		Dubs	1874	1930
2		Dubs	1875	1924
4a		Dubs	1877	1920
8		Dubs	1877	1919
13		Dubs	1883	1894
9	*Patience*	Cross	1880	1895
10	*Perseverance*	Cross	1880	1893
11		Vulcan	1880	1905

5		BP	1881	1940
6		BP	1881	1940
12		BP	1882	1925
16		BP	1890	1925
17		BP	1894	1935
18	Bantry	MW	1881	1906
14	St. Patrick	SS	1885	1908
15	St. Columba	SS	1885	1910
3		Dubs	1891	1930
9		Neilson	1895	1935
10		Dubs	1906	1933
18		Neilson	1894	1935
19		Baldwin	1900	1914
20		Baldwin	1900	1912
7		CB&SC	1901	1934
4	5954	BP	1919	1963
8	6034	BP	1906	1945
11	4752	BP	1919	1963
13	6077	BP	1920	1961
14	5265	B.P.	1909	1959
15	5413	BP	1910	1961
19	5822	BP	1914	1945
20	5616	BP	1912	1961
21a	Slaney 382	Hunslet	1885	1920
21b	St. Molaga 520	Hunslet	1890	1949
21c	Argadeen 611	Hunslet	1884	1957
?		Peckett	1920	1949

Locomotives

Three shades of olive and Brunswick green adorned these steam beasts with mid yellow and black lining. The carriage livery was the same but sometimes with cream lining.

Nos 1 & 2: were former Enfield steam carriages. They arrived in kit form during the summer of 1849. They had a 0-2-2 wheel arrangement (based on the size of the various wheels). There were no cabs on these engines which were named *Sighe Gaoithe* and *Rith Tineadh* so the poor fireman and driver were at the mercy of the elements.

No. 3: the *Fag an Beallach,* a Tayleur-built engine, arrived from England during September of 1849 with some damage owing to a rough sea journey across the Irish Sea. She broke an axle at Waterfall coincidentally at the very same spot that engine 207, an Ivatt 0-6-0, broke her axle in 1958. In 1964 a bright green C class engine broke down at the same location just under the footbridge.

No. 4: the cost of providing metal name plates had escalated to the point that it was decided to use only number plates on future locomotives. No. 4 was a sister locomotive to No. 3 and it had the same wheel arrangement. It was later sold in 1889.

Nos 5 & 6: Sharp Brothers delivered these two 0-4-2 heavy engines directly to Albert Quay by ship in October 1852. These engines proved troublesome and one was scrapped just twenty-two years later.

No. 7: this dark green saddle tank locomotive was a problematic yoke giving endless mechanical inconvenience on the Farrangalway section. She only lasted thirty seven years being cursed by driver and Fireman during her tenure.

No. 8: When second-hand locomotives were utilised on the Bandon line, No. 8, an ex-Great Southern & Western Railway's engine, was a Bury 0-4-2 and built in 1847. When she started on the Bandon line, she was already well past her sell-by-date and after many modifications was scrapped in 1872.

No. 9: here were two more clapped out Bury engines purchased from a grateful GS&WR but they didn't survive long, Suffice to say after this fiasco they then invested in five new Dubs built tank engines between 1874 and 1887 which gave great service and ran trouble free. Their numbers were 1, 2, 4, 8, and 13.

Nos 9b & 10: were former West Cork Company engines and their names were *Patience* and *Perseverance* and they survived until 1895.

No. 11: another ex-West Cork Company engine was a 2-4-0 saddle tank locomotive. It worked on until the twentieth century trouble free.

Nos 5, 6, 12, 16, 17: Beyer Peacock of (later Bandon tank fame) came with five classic 0-6-0 saddle tanks which ran trouble free from their 1881 birth to their 1940 rusty demise. Though not recognised as particularly handsome as their later sisters, they were as equally rugged.

Nos 14 & 15: were Rocksavage modified ex-Sharp Stewart built engines with a 4-4-0 wheel arrangement from 1893 and they suited the Ballymartle to Farrangalway section with its tight curves.

Nos 3 & 10: Dubs supplied two sturdy 2-4-0tank engines for goods trains in 1891 and 1893.

Nos 9 & 19 Neilson Reid built two tank engines for the Bandon line in 1894 which went on to give great service until scrapped by the cash strapped Great Southern Railways during the mid-1930s.

No. 20 at Rocksavage.

Nos 19 & 20: Burnham Parry & Williams & Co. provided the next two saddle tanks with an 0-6-2 wheel arrangement, built at the Balwin locomotive works, Philadelphia. If any locomotive didn't suit the damp Irish climate it certainly was this duo. They arrived at Albert Quay by the sea in 1900. They were tinny, troublesome and were thrashed by the outbreak of the Great War and were a waste of time and money and Chas T. Parry and Edward Williams were very red-faced over this débâcle. Yankee engines did make a comeback to Albert Quay over sixty years later and their only fault this time was their livery of drab black and dull tan.

No. 7: J. W. Johnstone oversaw the building of this engine at Rocksavage Works, Albert Quay station, in the summer of 1901, despite 15% of its parts being recycled from previous engines and the remaining 85% built with high grade materials. It was also renovated during the autumn of 1919 during the War of Independence and despite some damage done by an irate worker it survived until 1934.

No. 10: was a Rocksavage Works rebuild of a former Dubs 4-4-0 side tank. It later became a 4-6-0 tank, nick-named the Leaf Crusher. In 1925 its official Great Southern number was 471 with and it was scrapped in 1933.

Nos 3, 9, 18: were former Dubs/Neilson built 4-4-0 tank engines and were rebuilt at Rocksavage from 1898 to 1901 with lighter gauge metal.

Nos 4, 8, 11, 13, 14, 15, 19 &20: Beyer Peacock made the famous 4-6-0 tank locomotive, a handsome and rugged engine which could pull any size of a train up to Doughcloyne and Chetwynd and up to the summit at Waterfall. There was a children's book with Thomas the Tank style drawings of a puffing Bandon tank pulling a very heavy train up the steep Togher gradient saying 'I think I can! I think I can! I think I can! In a rhythmic chant changing to 'I knew I could! I knew I could! I knew I could!' having passed the Waterfall summit.

***

During the 1930s Berlin-based Orenstein and Koppel provided the water supply pumps. In the GSR days, engines 466, 469, and 470 were sent to Dublin to work the Bray section. When driving these engines, bunkers first, over Cork City Railway to Glanmire Road station, the driver would have to manoeuvre the rack lever with its razor sharp rachet to prevent a serious accident. These handsome engines never passed Skeaf on the Ballinascarthy to Courtmacsherry lines because of the very sharp curves. One of these engines was painted in the classic dark green Córas Iompair Éireann livery complete with beautiful *eau de nil* (light green) and black lining. Whilst black does not suit diesel or electric engines because you cannot see the windows, it certainly suited these beautiful Bandon tank locomotives. Engine 464 looked resplendent in black on the Irish Railway Record Society's West Cork Rail Tour on 17 March 1961.

When talking to old railwaymen years ago they called locomotive 463 'Kitty the Hare' because of her speed, No. 464 was called 'Henry' because of her Henry Ford style reliability, No. 470 was called 'The Grey Ghost' because of her quietness. I once went to a Purcell railway auction at Birr Offaly to bid on a Beyer Peacock Bandon tank makers plaque, one of many at auction that night. Unfortunately I was the unsuccessful underbidder on the night. There are now beautiful 4mm detailed scale models of these fine locos from professional model-making studios in Ennis. The Bandon tank was probably the finest and most attractive tank locomotive to operate in Ireland, or for that matter the United kingdom. Colm Creedon tried to preserve one during the CIÉ dark period 1962–1966, unfortunately without success.

The Cork Bandon South Coast Railway tank locos were handed over to the GSR, who then handed them over to CIE.

CBSCR	CIE	Built	Rebuilt	Withdrawn
No. 4	No. 463	1919	1950	1963
No. 8	No 464	1920	1946	1963
No. 11	No. 465	1906		1945
No. 13	No. 466	1920	1947	1961
No. 14	No. 467	1909	1935	1959
No. 15	No. 468	1910	1948	1961
No. 19	No. 469	1914		1945
No. 20	No. 470	1912	1948	1961

Following the 1925 GSR takeover, steam engines swarmed across the West Cork rail system. GSR Nos 30, 31, 32, 33, 34. 35, 36, 37, 41, 90, 100, 201, 207, 208, 217, 269, 270, 274, 299, 417, 423, 432, 433, 459, 490, 495, 551, 552, 556, 557, plus at least a half a dozen unrecorded tank locos were in use up to the closure. During 1960 Ray Good worked at Clonakilty junction and noted that every Metro-Vick C class diesel loco has operated on the West Cork rail system. A class diesel engines have travelled on trials down the line. E class diesels have travelled down to Courtmacsherry to determine their quality, safety, performance and usefulness.

The first livery was silver with Tóstal green lettering. The second was everyone's favourite – a bright green cheerful colour with *eau-di-nil* stripe and lettering, a design classic. From 1954 various AEC Park Royal railcar sets operated on all the West Cork routes with the exception of the Courtmacsherry branch. No. 2637 looked beautiful with its dark Brunswick green colour but railcar No. 2641 stole the show with her attractive bright Tóstal green.

These trains were highly praised by the travelling public from Bantry to Belfast and beyond. The seats behind the driver put you in visual touch with nature as you had full frontal and side screen views because of the very sizeable windscreen and large side windows. In comparison the old steam engines had very small dinner plate sized windows. Driver Jack Daly once stated a 'Raneen' couldn't look out these tiny windows. A 'Raneen' is a West Cork/ East Kerry saying to describe a small bird such as a wren.

After 1969 General Motors diesel engines also performed some sterling service to and from Albert Quay. They were not good-looking like the A or C class but they were far more reliable. A typical Yankee road switcher: rough, gruff, tough and couldn't get enough.

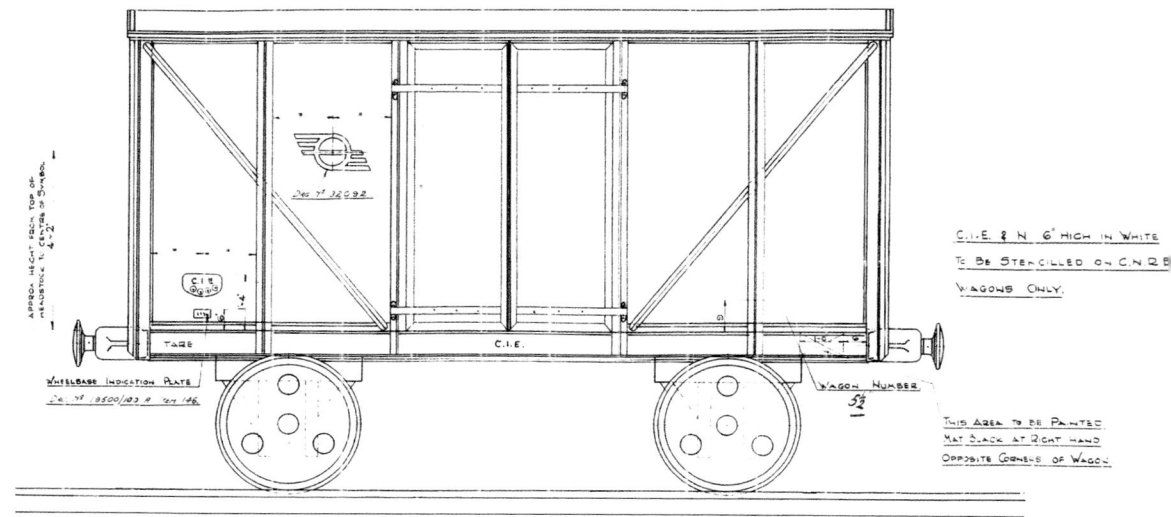

Goods van.

Wagons

Year	Covered Number	Box Wagon	10 & 20 ton open	Cattle Wagon	Brake
1851-1902	60, 61, 65 66, 71, 78, 84, 95, 106, 111, 119, 121, 123, 126, 128, 137, 141, 142, 145, 176, 177, 193, 197, 229, 235, 238, 241, 243, 245, 254, 254, 255, 257, 258, 259, 260, 262, 267, 269, 270, 278, 279, 280, 281, 283, 286, 288, 290, 291, 292, 295, 297, 313, 317, 319, 320, 322, 323, 324, 324, 325, 327, 328, 329, 331, 332, 342, 344, 345, 346, 347, 350, 352, 355, 356, 358, 360	67, 138, 194, 240, 248, 265, 336		403, 405, 407, 416, 417, 420, 421, 423, 424, 426, 427	756, 759
1902/03 - 1904			557, 558, 559, 560, 561, 562, 563, 564, 565, 566, 567, 568, 572, 573, 574, 575, 576, 577, 578, 579, 580, 581, 552, 583		

Year	Covered Number	10&20 ton open	Cattle Wagon	Brake
1905		551, 552, 553 554, 555, 556	401, 402, 404 406, 408, 409	
1906	1, 2, 3, 4		410, 411, 412, 413	
1907	5, 6	584, 585, 586, 587, 588		
1908	68, 104, 138, 233, 237, 284, 289, 293, 294, 296, 318, 326, 337, 340, 349	589, 590, 591, 592, 593, 594		
1909	38, 58, 108, 115, 130, 209	595, 596, 597, 598, 599, 600		
1910		601, 602, 503, 604, 605, 606, 607, 608, 609, 610, 611, 612		761, 762
1911		613, 614, 615, 616, 618, 619, 620	428, 429, 430, 431, 432, 433	
1912	7, 8, 9, 10, 11, 12, 13, 14, 15, 715, 434, 435, 436, 715, 718		434, 435, 4 36	
1913	17, 18, 92, 59		34, 42, 48, 50, 70	763
1914	59		43, 51, 52 79, 80, 81, 82, 83, 86, 118	

Year	Covered Number	10&20 ton open	Cattle Wagon	Brake
1915	19a (6 wheel), 19 (4 Wheel) 751, 757	338	85, 91, 93, 175, 239	760
1916		621,622,623 624,626,627, 629,630	96, 97, 144, 200, 211, 223	
1917			36,53,54, 64, 69, 77, 87,90,99, 100,103	
1918			45,62,107, 114,122	
1919	124,125,129		72,131, 133,134	
1920			55,74,132,143, 147, 148,151, 154,155,156	
1921		635,636,637 638, 639, 640, 641 642, 643,644 645, 646, 647, 648, 649, 650	159,160,161, 162,163, 164,165, 166,167	Damaged during Civil War
1922			41, 88,98,105, 135,152	
1923			56,102,116	
1924			35,44,46,47	

At least a dozen of these wagons both covered and uncovered were damaged during the Civil War.

When the Great Southern railways began to operate from Albert Quay in 1925, they scrapped quite a few Bandon wagons and brought in many of their own G.S.R. wagons known as four wheeled mongrels. Very large white G.S. letters replaced the former cream coloured C.B. letters. One brake van even had the Cork Bandon crest painted on her side. The Rocksavage painters left this crest on the brake van until she was broken up during the Second World War. Colm Creedon informed me that there are unrecorded gaps regarding Bandon wagon numbers, he was unable to plug the gaps. One character called Buck Stoat Billy used to carry poitín on these brake vans for resale in the northside of the city. Guns were planked (hidden) in the brake van stoves during the Civil War fooling the Free State troops who were guarding Albert Quay Station night and day.

Year	Brake Van	Withdrawn	Colour	Number
1896	Brake/Passenger	1948	Black	71
1896	Brake/Passenger	1960	Black	72
1898	Brake/Passenger	1940	Black	73
1899	Brake/Passenger	1957	Black	74
1922	Brake/Passenger	1967	Grey	75
1902	Brake/Passenger	1957	Grey	76
1902	Brake/Passenger	1911	Black	77
1902	Brake/Passenger	1911	Black	78

Year	Horsebox	Withdrawn	Colour	Number
1878	Horsebox	1908	Black	1
1896	Horsebox	1923	Grey	2
1898	Horsebox	1949	Green	3
1899	Horsebox	1949	Green	4

Year	Fish Van	Withdrawn	Colour	Number
1899	Fish Van	1923	Dark Grey	1
1899	Fish Van	1949	Black	2

| 1899 | Fish Van | 1949 | Dark Grey | 3 |

Year	Carriage Truck	Withdrawn	Colour	Number
1871	Carriage Truck	1962	Green	4
1890	Carriage Truck	1941	Grey	5

Macroom Section Stations	Original Colours	Colours
Bishopstown	Wine and White	Green and Cream
Ballincollig	Wine and White	–
Kilumney	Black and Cream	Green and Cream
Kilcrea	Wine and White	Green and Cream
Crookstown Rd	Black and Cream	Green and Cream
Dooniskey	Wine and White	Green and Cream
Macroom	Wine and White	Green and Cream

On this line the gatekeepers chose their own colour schemes for the various houses on the line as did Bandon after 1955. Some Great Southern enamel name boards were not put up on the Macroom section thus leaving some of the original signs in place until closure. On the Bandon system the black and white enamel name board signs replaced the earlier ones with the exceptions of Creagh, Baltimore, Clonakilty, Skibbereen, Aughaville, Ballymartle and Farrangalway. The Schull and Skibbereen narrow gauge stations were given the enamel signs in August 1925. Every name board on the West Cork system has been preserved with the notable exception of Bandon. The vast majority of these signs are in the homes of UK rail fans whilst the rest are on the walls of Irish bars in New York.

APPENDIX 5

LEVEL CROSSINGS

Cork to Bantry
(Signals are only needed where there are stations, junctions and tunnels)

Place	Distance	Signals
Twomey's Castlewhite	5 miles 474 yards	no
Gortnaclough	12 miles 409 yards	signal
Killeen	13 miles 1,165 yards	no
Lisnagroom	13 miles 1,595 yards	no
Lisiniskery	14 miles 600 yards	no
Dunkereen	14 miles 1,160 yards	no
Crosses	15 miles 554 yards	signal
Rockfort	15 miles 1.456 yards	signal
Chapel	20 miles 680 yards	no
Rice's Road	20 miles 100 yards	no
Castlebernard	21 miles 640 yards	no
Enniskeane	29 miles 1,264 yards	signal
Nedinagh	34 miles 1,355 yards	no
Ballyboy	36 miles 505 yards	signal
Millenananig	36 miles 1,200 yards	no
Station Gates	42 miles 425 yards	signal
Lougherot	42 miles 1,424 yards	no
Derrynagree	44 miles 170 yards	down
Drimoleague	45 miles 115 yards	no
Station Gates	45 miles 1,530 yards	signal
Bog No. 1	46 miles 490 yards	no
Bog No. 2	46 miles 1,650 yards	no
Inchingerig	48 miles 175 yards	no

Station Gates	49 miles 800 yards	signal
Station Gates	51 miles 500 yards	no
Keilnascarta	54 miles 565 yards	no
Gurtnamuck	57 miles 310 yards	down
Old Barrack Road	57 miles 1,650 yards	signal
Station Gates 1	58 miles 15 yards	signal
Station Gates 2	58 miles 790 yards	signal

Clonakilty to Clonakilty junction

Place	Distance	Signals
Cashelmore	26 miles 260 yards	signal
Ahalisky	28 miles 1,110 yards	no
County Home	32 miles 986 yards	no

Ballinascarthy to Courtmacsherry

Place	Distance	Signals
Ballinascarthy	0 miles 390 yards	signal
Monteen	1 mile 490 yards	no
Inchy Bridge	4 miles 500 yards	up
Station Gates 1	5 miles 1,670 yards	no
Station Gates 2	6 miles 145 yards	no

Drimoleague junction to Baltimore

Place	Distance	Signals
Station Gates	45 miles 1,364 yards	signal
Garranes South	46 miles 870 yards	no
Reenroe	48 miles 1,000 yards	signal
Station Gates	49 miles 523 yards	signal
Cooragannive	49 miles 1,684 yards	no
Back of town	53 miles 1,353 yards	signal

Place	Distance	Signals
Skibbereen Street	53 miles 1,565 yards	signal
Coronea No. 1	54 miles 740 yards	no
Coronea No. 2	54 miles 1,200 yards	signal
Mallavonea	55 miles 550 yards	no
Bunlick	56 miles 480 yards	up
Creagh No. 1	57 miles 575 yards	signal
Creagh No. 2	57 miles 1,320 yards	no
Rath No. 1	59 miles 820 yards	signal
Rath No. 2	59 miles 1,135 yards	signal
Church Strand	60 miles 880 yards	signal
School Crossing	61 miles 909 yards	no

Cork and Macroom Section

Place	Distance	Signals
Ballincollig Gates	5 miles 1,555 yards	signal
Grange Gates	8 miles 560 yards	down
Station Gates	8 miles 1230 yards	no
Ballast Crossing	8 miles 1,465 yards	no
Kilcrea	11 miles 0 yards	no
Station Gates	12 miles 140 yards	signal
Coolmucky	14 miles 645 yards	no
Station Gates 1	16 miles 0 yards	signal
Station Gates 2	19 miles 625 yards	signal
Station Gates 3	23 miles 1,140 yards	signal

There were forty-seven cottages on the system falling to forty-one after the closure of the Cork to Macroom line. Many a family were born and reared in these well-kept and tended homes. For obvious reasons terms and conditions applied to each and every halt keeper irrespective of whether they were male or female. Sometimes a wife would be the halt keeper while her husband could work on one of the many other jobs on the West Cork system. The various railway companies down through the years frowned upon heavy drinking halt keepers, as the safety aspects if ignored or flaunted could create an extremely dangerous situation. Their responsibilities were no less that those of the signalmen, engine drivers and firemen.

Dogs had to be trained to cope with trains and humans at the level crossing area. Cats, hens, ducks and geese also became part of this living railway harmony.

Milepost 8 Courtmacsherry

APPENDIX 6

STATIONS

Cork to Bantry

Miles	Station	Opened	Closed
*0-0	Albert Quay	1851	1976
3-10	Doughcloyne Halt	1851	1853
4-40	Meagher's Lane Halt	1851	1853
8-35	Waterfall	1852	1961
9-26	Ballinhassig	1849	1961
13-20	Kinsale Junc./Crossbarry	1863	1961
15-35	Innishannon Rd/Upton	1849	1961
19-65	Bandon Station	1849	1961
21-50	Castle Bernard	1873	1961
24-00	Clonakilty Junction	1886	1961
27-64	Desert	1867	1961
30-00	Ballineen/Enniskeane	1891	1961
34-15	Manch Halt	1868	1891
34-25	Dunmanway Shed	1866	1877
37-56	Dunmanway	1877	1961
38-00	Atkin's Hall (Goods)	1890	1961
42-00	Knockbue Halt	1878	1961
45-50	Drimoleague	1866	1877
45-60	Drimoleague Junction	1877	1961
49-44	Aughaville Halt	1886	1961
51-78	Durrus Road Halt	1881	1892
57-20	Bantry (Hilltop)	1881	1892
57-60	Bantry (Town)	1892	1961
58-60	Bantry (Pier)	1892	1949

CORK TO BANDON AND BANTRY.

Distance from Cork.	DOWN TRAINS	Sectional Running Pas.		Sectional Running Gds.		1. arr.	1. dep.	2. arr.	2. dep.	3. Goods D.E. arr.	3. Goods D.E. dep.	4. arr.	4. dep.	5. PAS. D.T. arr.	5. PAS. D.T. dep.	6. PAS. D.T. arr.	6. PAS. D.T. dep.	7. arr.	7. dep.
		D	S	D	S														
Mls.						a.m.	a.m.	a.m.	a.m.	a.m.	a.m.	a.m.	a.m.	p.m.	p.m.	p.m.	p.m.	p.m.	p.m.
—	CORK (Albert Quay) W ¶ ●	0	0	0	0	...	...	...	3 45	...	7 00	...	...	...	12 15	...	6 00	...	...
6½	WATERFALL HALT ¶	14	15	24	28	...	...	...	4 11	...	7 26	...	...	12 30	12 31	6 15	6 16	...	...
10	BALLINHASSIG HALT ¶	7	9	10	10	...	...	...	4 21	...	7 36	...	...	C.R.		6 24	6 25	...	...
13¼	CROSSBARRY H W ¶	5	8	9	10	...	...	4 32	4 33	7 47	7 54	...	...	12 44	12 45	6 31	6 32	...	...
15½	UPTON HALT + ¶	3	4	6	7	...	...	...	4 41	C.R.	8 04	...	...	12 49	12 50	6 36	6 37	...	...
20	BANDON ... W ● ¶	6	8	9	9	...	...	4 52	4 54	8 17	9 20	...	...	12 57	12 59	6 44	6 47	...	...
24	CLONAKILTY JUNC. ¶ HALT	7	6	10	11	...	...	5 08	8 18	9 34	9 49	...	...	1 07	1 08	6 55	6 57	...	...
27¾	DESERT HALT ...	5	6	7	8	...	...	...	5 27	10 00	10 10	...	...	1 14	1 15	7 03	7 04	...	...
30	BALLINEEN ¶W	3	4	5	6	...	...	5 34	5 38	10 19	10 47	...	...	1 19	1 20	7 08	7 09	...	...
37¾	DUNMANWAY ¶W	10	12	14	17	...	...	5 56	6 00	11 05	11 26	Diesel Light Engine p.m. Noon ... 12 00		1 31	1 32	7 20	7 21	...	...
42	KNOCKBUE HALT	...	...	...	...	...	...	...	...	...	...			...	...	...	...	...	...
45½	DRIMOLEAGUE ● ¶W	12	14	20	24	...	...	6 24	7 00	11 30	...			1 45	1 46	7 34	7 38	...	...
...	AUGHAVILLE H ...	...	...	...	...	...	...	...	...	...	...	...	...	C.R.		C.R.		...	...
52	DURRUS ROAD H N	13	19	17	22	...	...	7 21	7 26	...	...	...	...	C.R.		C.R.		...	...
57½	BANTRY ● ¶W	10	16	18	18	...	...	7 50	...	...	...	12 25	...	2 10	...	8 05	...	...	...

—— Indicates where trains cross or pass each other.   N.—No Telephone Communication.

C.R.—Will stop when required.

Cork to Bantry timetable.

BANTRY AND BANDON TO CORK.

Distance from Baltimore.	UP TRAINS	Sectional Running				WEEK-DAYS										
						8. Light Engine D.E.		9.		10. PAS. D.T.		11.		12. PAS. D.T.		13. Goods D.E.		14. PAS.	
		Pas.		Gds.		arr.	dep.	arr.	dep.	arr.	dep.	arr.	dep.	arr.	dep.	arr.	dep.	arr.	dep.
		D	S	D	S	a.m.	a.m.			a.m.	a.m.			p.m.	p.m.	p.m.	p.m.		
	BANTRY ●W	0	0	0	0	...	8 55	...	...	...	8 25	...	...	...	3 00	...	12 45	...	...
	DURRUS ROAD H N	14	16	23	25	...	...	...	...	C.R.		...	...	C.R.		C.R.	1 10	...	...
	AUGHAVILLE H	...	...	...	...	...	...	...	...	C.R.		...	...	C.R.		...	...	...	...
16¼	DRIMOLEAGUE W ●	12	19	19	22	9 25	9 40 Gds.	...	...	8 52	8 57	...	...	3 27	3 29	1 31	4 40	...	...
19¾	KNOCKBUE HALT	...	...	...	...	...	...	...	...	...	...	...	...	...	...	...	...	...	...
24	DUNMANWAY W	13	15	23	26	10 07	10 27	...	...	9 11	9 12	...	...	3 43	3 44	5 07	5 12	...	...
31¼	BALLINEEN W	10	15	14	16	10 45	11 10	...	...	9 23	9 24	...	...	3 55	3 56	5 30	5 40	...	...
34	DESERT HALT	3	4	4	4	11 18	11 23	...	...	9 28	9 29	...	...	4 00	4 01	...	5 46	...	...
37½	CLONAKILTY ● JUNCTION HALT	6	6	7	7	11 34	11 50	...	...	9 36	9 38	...	...	4 08	4 09	5 55	6 10	...	...
41¼	BANDON W	7	6	8	8	12 02	1 00	...	...	9 46	9 48	...	...	4 17	4 18	6 22	6 55	...	...
46¼	UPTON HALT +	6	8	12	13	1 16	1 17	...	...	9 55	9 56	...	...	4 25	4 26	...	7 09	...	...
48¼	CROSSBARRY H W	3	4	6	6	1 27	1 28	...	...	10 00	10 01	...	...	4 30	4 31	7 17	7 18	...	...
51¾	BALLINHASSIG HALT	6	8	15	20	1 47	1 48	...	...	10 08	10 09	...	...	C.R.		...	7 35	...	...
55¼	WATERFALL HALT	7	9	16	21	2 08	2 09	...	...	10 17	10 20	...	...	4 45	4 48	...	7 51	...	...
61¼	CORK (Albert Quay) W ●	11	14	17	18	2 30	...	...	...	10 35	...	...	...	5 00	...	8 10	...	...	...

——— Indicates where trains cross or pass each other.   N—No Telephone Communication.

C.R.—Will stop when required.

Bantry and Bandon to Cork.

Kinsale Junction to Kinsale

13-20	Kinsale Junc./Crossbarry	1863	1961*
17-03	Ballymartle	1863	1931
21-07	Farrangalway	1863	1931
24-00	Kinsale	1863	1931

Clonakilty Junction to Courtmacsherry

24-00	Clonakilty Junction	1886	1961
29-20	Ballinascarthy Junction	1886	1961
31-00	Shannonvale (Goods)	1887	1961
33-00	Clonakilty	1886	1961
00-00	Ballinascarthy Junction	1886	1961
3-00	Skeaf	1890	1961*
5-00	Umerra Stone Quarry	1890	1938
5-78	Timoleague	1890	1961
9-10	Courtmacsherry	1891	1961

* Mileposts from Cork
*Used by the Larkin and Deasy families until its closure

Drimoleague Junction to Baltimore

45-52	Drimoleague Junction	1877	1961
48-69	Madore	1877	1961
53-49	Skibbereen	1877	1961
57-30	Creagh	1893	1961
60-21	Baltimore	1893	1961

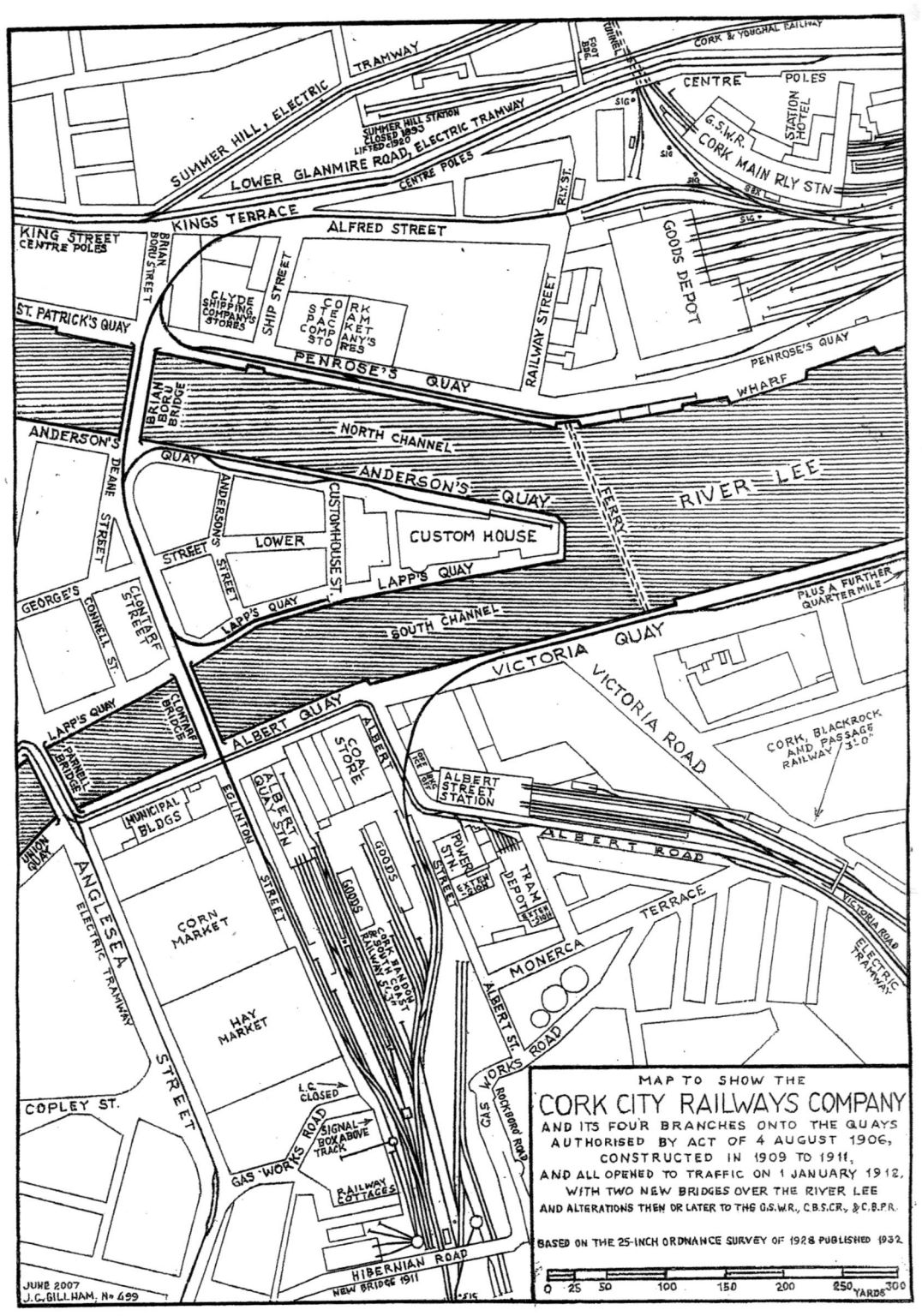

Map of Cork City Railways Company.

CORK CITY RAILWAYS

Miles	Station	Opened	Closed
0-15	Glanmire Road	1912	1976
0-60	Albert Quay	1912	1976

All the station staffs have been preserved from these various sections, most of these are now in Germany and the UK. The late P. J. O'Meara (a famous railway enthusiast) had a huge collection of railway staffs, numbering over 600, and they were all sold at Purcell's auction a few years ago. Many were snapped up by internet buyers and are now scattered all over the globe.

An RIC officer waits for the Baltimore train to pass at the Blackrock Road Rock cutting, 24 July 1914.

Opening and Closing Dates of West Cork Railway Sections

Opening and Closing Dates	Opened	Closed
Ballinhassig to Bandon	1 August 1849	1 April 1961
Albert Quay to Ballinhassig	8 December 1851	1 April 1961
Bandon to Dunmanway	12 June 1866	1 April 1961
Dunmanway to Skibbereen	23 July 1877	1 April 1961
Kinsale Junction to Kinsale	27 June 1863	1 September 1931
Clonakilty Junction to Clonakilty	28 August 1866	1 April 1961
Ballinascarthy to Timoleague	30 December 1890	1 April 1961
Timoleague to Courtmacsherry	29 April 1891	1 April 1961
Drimoleague to Bantry	4 July 1881	1 April 1961
Bantry to Bantry Town	22 October 1892	1 April 1961
Bantry Town to Bantry Pier	2 June 1909	1 September 1937
Skibbereen to Baltimore	2 May 1893	1 April 1961
Cork to Macroom	12 May 1866	10 Nov 1953
Macroom Junction to Cork	27 September 1879	2 March 1925

Schull and Skibbereen Narrow-gauge railway

Opening and Closing Dates	Opened	Closed
Skibbereen to Schull	9 September 1886	27 January 1947

Near Woodlands, facing Schull in 1938.

Crooked Bridge in 1938.

Chris Larkin in original railway apparel.

ACKNOWLEDGMENTS

I have been very fortunate in writing this book as direct historical information and data has been passed down from the original railway generation of 1849 to my grandfather's generation of the 1890s, down to my uncle's generation of the 1950s.

I also owe a depth of gratitude to my brother Jerry, who was there to help in very way he could; Michael Lenihan, for his wise support and his typing of my original handwritten manuscript; David Bate for his constant photographic support and advice; Liam Murphy of Mishells who scanned the images in a professional way.

Thanks are also due to the West Cork railway experts: Walter McGrath of the Evening Echo, Colm Creedon, the St Leger brothers – Joe and John – and Ray Good of the IRRS whose vast and gigantic archives were always available to me.; Ciarán Cooney and Tim Moriarty of IRRS headquarters were helpful in every way.

Further thanks go to Mary Lucey of Bandon Office Supplies; Alan and Leslie Hyland of the IRRS; Fr Tom Davitt of the IRRS; Una O'Donoghue and Arthur Lawton, John O'Donoghue (RIP) of the Ford motor company; Hugh McCarthy, Manchester; Michael Larkin of Kilbrittan; Thomas O'Keeffe; Leo McMahon, *Southern Star*; Peter Harding photos , Dick Fearns, Conor Clune, Eddie O'Connor and Geraldine Finucane of Iarnród Éireann. If I've omitted anybody through forgetfulness, I sincerely apologise.

A special word of thanks to all at Mercier Press.

Photo acknowledgements: Thanks are due to the following very helpful people for supplying images: N. McAdams, 52; S. O'Brien 46, 78, 89, 164; D. Bate, 89, 101, 102, 104, top p 106, 107, 108, 116, top 104, top 105, 133, 134, 157, 159, 162, 163; CIE, 120, 121; C. Creedon, 9, 44, 74, 85, 149, 153, 163, 170; CD. Gammell, 4; W. McGrath, 13, 16, 24, 26, 32, 40, 42, 69, 70, 80, botton 140, 152, 201, 221; P. Harding, 54, 222; L. Hyland, 73; J. St Leger, 20, 45, 136; J. Langford, top 88; IRRS, 66, 68, 93, 94, bottom 119, 126, bottom 142; D. Soggue, 30; the rest of the images are from my collection.

MERCIER PRESS
Cork
www.mercierpress.ie

© Chris Larkin, 2022

ISBN: 978 1 78117 776 1

Cover design and typesetting: Sarah O'Flaherty

A CIP record for this title is available from the British Library.

This book is sold subject to the condition that it shall not, by way of trade or otherwise, be lent, resold, hired out or otherwise circulated without the publisher's prior consent in any form of binding or cover other than that in which it is published and without a similar condition including this condition being imposed on the subsequent purchaser.

No part of this publication may be reproduced or transmitted in any form or by any means, electronic or mechanical, including photocopying, recording or any information or retrieval system, without the prior permission of the publisher in writing.

Printed and bound in the EU.